SpringerBriefs in Population Studies

Population Studies of Japan

The world population is expected to expand by 39.4% to 9.6 billion in 2060 (UN World Population Prospects, revised 2010). Meanwhile, Japan is expected to see its population contract by nearly one third to 86.7 million, and its proportion of the elderly (65 years of age and over) will account for no less than 39.9% (National Institute of Population and Social Security Research in Japan, Population Projections for Japan 2012). Japan has entered the post-demographic transitional phase and will be the fastest-shrinking country in the world, followed by former Eastern bloc nations, leading other Asian countries that are experiencing drastic changes.

A declining population that is rapidly aging impacts a country's economic growth, labor market, pensions, taxation, health care, and housing. The social structure and geographical distribution in the country will drastically change, and short-term as well as long-term solutions for economic and social consequences of this trend will be required.

This series aims to draw attention to Japan's entering the post-demographic transition phase and to present cutting-edge research in Japanese population studies. It will include compact monographs under the editorial supervision of the Population Association of Japan (PAJ).

The PAJ was established in 1948 and organizes researchers with a wide range of interests in population studies of Japan. The major fields are (1) population structure and aging; (2) migration, urbanization, and distribution; (3) fertility; (4) mortality and morbidity; (5) nuptiality, family, and households; (6) labor force and unemployment; (7) population projection and population policy (including family planning); and (8) historical demography. Since 1978, the PAJ has been publishing the academic journal *Jinkogaku Kenkyu* (The Journal of Population Studies), in which most of the articles are written in Japanese.

Thus, the scope of this series spans the entire field of population issues in Japan, impacts on socioeconomic change, and implications for policy measures. It includes population aging, fertility and family formation, household structures, population health, mortality, human geography and regional population, and comparative studies with other countries.

This series will be of great interest to a wide range of researchers in other countries confronting a post-demographic transition stage, demographers, population geographers, sociologists, economists, political scientists, health researchers, and practitioners across a broad spectrum of social sciences.

Yoshitaka Ishikawa
Editor

Japanese Population Geographies I

Migration, Urban Areas, and a New Concept

 Springer

Editor
Yoshitaka Ishikawa
Professor Emeritus
Kyoto University
Kyoto, Japan

ISSN 2211-3215 ISSN 2211-3223 (electronic)
SpringerBriefs in Population Studies
ISSN 2198-2724 ISSN 2198-2732 (electronic)
Population Studies of Japan
ISBN 978-981-99-2034-1 ISBN 978-981-99-2035-8 (eBook)
https://doi.org/10.1007/978-981-99-2035-8

This Springer imprint is published by the registered company Springer Nature Singapore Pte Ltd.
The registered company address is: 152 Beach Road, #21-01/04 Gateway East, Singapore 189721,
Singapore

Preface

According to the World Population Prospects 2022 issued by the Population Division, United Nations, the population growth rate in Japan in 2021 was −0.54%. Among the 38 OECD countries, a negative rate has also been recorded for 11 countries (Czechia, Estonia, Greece, Hungary, Italy, Latvia, Lithuania, Poland, Portugal, Republic of Korea, and Slovakia); Japan's rate of decrease follows those of Latvia (−1.38%), Lithuania (−1.29%), and Greece (−0.69%). However, if only countries with a population of more than 50 million were counted, Japan would have the highest rate of decrease. Therefore, Japan is a representative population-decreasing country in the contemporary developed world. Furthermore, the COVID-19 pandemic from 2020 has caused a sudden decline in total fertility rates (TFR) throughout the world. If the pandemic further accelerates the decline of TFR, the era of depopulation will come sooner than expected. This means that a rapidly increasing number of countries will have to deal with the risk of population decline as an urgent issue.

Over the past decade, the term *jinko gensho* (population decline) has become a popular expression in Japan: The total population of Japan began to drop after reaching its peak (128.08 million) in 2008. Following the long-term population projections by Japan's National Institute of Population and Social Security Research (IPSS), showing that the country's annual decrease will gradually speed up, this issue's seriousness has become widely recognized, prompting myriad discussions on population decline.

It goes without saying that this demographic trend is a painful experience for Japan. Nevertheless, the recent population-related trends observed in Japan should provide many valuable lessons for countries around the world that are in the midst of population decline or facing its imminent emergence. Consequently, at this juncture we have an obligation to let other countries know about the achievements in population geographic studies in Japan. Existing research in Japan has received substantial stimuli from foreign research, especially from English-speaking countries. Unfortunately, however, Japanese population geographers are reluctant to publish their findings in English, often publishing their work only in Japanese journals. Thus, even papers with excellent findings and insights are not widely known around the world.

Therefore, the editor came up with the idea of compiling an anthology of English translations of recently published Japanese papers that are regarded as important achievements in Japanese population geography. Given the rapid progress of automatic translation in recent years, this may be the last chance to publish a fully translated anthology, and it's possible that similar projects will not be carried out in the future. It should be noted, however, that not all recent achievements in population geography in Japan deal with issues related to population decline, and interest in them varies. Accordingly, the relationship to population decline in each of the anthology's papers is explicitly described in its introduction below.

This work is published by the Population Association of Japan (PAJ) as a *Population Studies of Japan* Series through Springer. According to the agreement between PAJ and Springer, the length of a single book is limited to 125 pages, which is insufficient to also include the most recent five papers. For this reason, it was decided to publish a total of ten papers in two volumes, each containing five papers.

The two English volumes comprise the following original papers published in Japanese.

The first volume (Ishikawa Y (ed) (2023) *Japanese Population Geographies I: Migration, Urban Areas and a New Concept*) includes the following five chapters:

Chapter 1: Inoue T (2016) *Posuto jinko tenkan-ki no jinko ido* (Internal Migration in the Post-Demographic Transition Period). In: Sato R., Kaneko R. (eds) *Posuto jinko tenkan-ki no nihon* (Japan in the Period of Post-demographic Transition), Hara Shobo, Tokyo, 111–133.

Chapter 2: Ishikawa Y (2016) *Nihon no kokunai intai ido saiko* (Internal Retirement Migration in Japan Revisited). In Ishikawa Y. (ed) *Ryunyu gaikokujin to nihon: Jinko gensho no shohosen* (New Immigration and Japan: Solution to Population Decline), Kaisei-sha, Otsu, 119–146.

Chapter 3: Yamada H (2020) *Higashi-nihon daishinsai no hisaichi ni okeru kyojuchi ido to shigaichi saihen tono kankei: Tohoku chiho no hisai-ken ni chakumoku shite* (Characteristics of Residential Mobility After the Great East Japan Earthquake: Focusing on Affected Prefectures of the Tohoku Region, Japan). *Kikan Chirigaku* (Quarterly Journal of Geography) 72(2): 71–90. https://doi.org/10.5190/tga.72.2_71

Chapter 4: Kanda H, Isoda Y, Nakaya T (2020) *Jinko gensho kyokumen ni okeru nihon no toshi kozo no hensen* (Spatial-Cycle Model Phases and Differential Urbanization of Cities in the Era of National Population Decline: Japanese Cities 1980–2015). *Kikan Chirigaku* (Quarterly Journal of Geography) 72(2): 91–106. https://doi.org/10.5190/tga.72.2_91

Chapter 5: Sakuno H (2019) *Jinko gensho shakai ni okeru kankei jinko no igi to kanosei* (Significance and Possibilities of the New Concept of "Relationship Population" in Japan's Population Decline Society). *Keizai Chirigaku Nenpo* (Annals of the Association of Economic Geographers) 65(1): 10–28. https://doi.org/10.20592/jaeg.65.1_10

The second volume (Ishikawa Y (ed) (2023) *Japanese Population Geographies II: Minority Populations and Future Prospects*) includes the following five chapters:

Chapter 1: Takeshita S, Hanaoka K, Ishikawa Y (2020) *Hetero-rokarizumu-ron no kensho: Aichi ken no toruko-jin no kyoju patan ni shoten o atete* (Investigating

Empirical Validity of Heterolocalism: Focusing on Turkish Residential Patterns in Aichi Prefecture). *Aichi Gakuin Daigaku Bungakubu Kiyo* (Bulletin of the Faculty of Letters of Aichi Gakuin University) 50: 65–74

Chapter 2: Yamauchi M (2021) *Osaka-shi ni okeru seiteki mainoritei no kukan bunpu* (Examining Geographic Distribution of LGBTs in Osaka City, Japan). *Jinko Mondai Kenkyu* (Journal of Population Problems) 77(2): 185–205. https://www.ipss.go.jp/syoushika/bunken/data/pdf/21770207.pdf

Chapter 3: Koike S (2021) *Nihon no chiiki-betsu shorai jinko no mitooshi* (Future Prospects of Regional Population in Japan). *Jinko Mondai Kenkyu* (Journal of Population Problems) 77(2): 85–100. https://www.ipss.go.jp/syoushika/bunken/data/pdf/21770201.pdf

Chapter 4: Tanimoto R (2017) *Toshi kogai ni okeru byosho heno akuseshibiritei no shorai suikei: Osaka toshiken hokubu no jirei* (Future Projection of Accessibility to Hospital Beds in the Suburbs: The Case of the Northern Osaka Metropolitan Area). *Jimbun Chiri* (Japanese Journal of Human Geography) 69(4): 425–446. https://doi.org/10.4200/jjhg.69.04_425

Chapter 5: Nakazawa T (2018) *Seiji keizaiteki jinko chirigaku no kanosei: "Shukusho nihon no shogeki" o tegakari ni* (Toward a Politico-Economic Population Geography: A Critique of *The Shock of a Shrinking Japan*). *Keizai Chirigaku Nenpo* (Annals of the Association of Economic Geographers) 64(3):165–180. https://doi.org/10.20592/jaeg.64.3_165

These 10 papers were originally published in Japanese, and for them to be included in this series, the editor had to ask the authors of each chapter to shorten them due to the length limitation mentioned above. Thus, several of the English translations are shorter than the original papers in Japanese, with some of them shorter by almost half. Consequently, in those substantially truncated English papers some of the excellent information from the original Japanese versions had to be omitted.

The following gives a brief overview of the significance of each paper in this series. The first volume focuses on internal migration, which has been the central theme of population geographic studies in Japan, as well as the structural changes in metropolitan areas, which are closely related to internal migration. It also includes a paper discussing a new population concept originating from rural areas in peripheral regions where population decline is most severe.

Inoue's paper in Chap. 1 analyzes the internal migration in post-war Japan and examines the influence of the demographic transition on so-called migration turnarounds in the post-demographic transition period. The author is one of Japan's leading population geographers and has many achievements in this field. Changes in migration between the three major metropolitan areas of Tokyo, Osaka, and Nagoya, which are the country's core regions, and their peripheral regions have been the focus of much attention in studies on population geography in Japan. The author discusses when and how the balance changed between the mainstream migration from the periphery to the core and the counterstream migration from the core to the periphery, as well the factors behind this change. Focusing on the migration turnarounds that have been observed twice in post-war Japan, the author carried out a wide-ranging study. The measure called "cohort cumulative social increase ratio," proposed by the author through his research, is particularly useful in studies that delve into the Japanese characteristics mentioned above.

Ishikawa's paper in Chap. 2 examines retirement migration, which has long attracted considerable attention in population geographic research in Japan. The predominant view is that retirement migration triggered by mandatory retirement has been widely observed in Western countries but not commonly in Japan. However, using the data from the 2010 Census, the author conducted a comprehensive analysis by creating a migration schedule for all prefectures and municipalities. Since retirement migration tends to be directed to rural areas where living costs are low, the influx of retired migrants is good news for municipalities in rural areas with marked population decline. From his analysis, the author confirms remarkable human flows to many prefectures/municipalities in peripheral Japan; this result indeed provides positive evidence of retirement migration. He also mentions the regional differences and assessments of retirement migration by officials of local governments in the major destinations of this migration. The paper mainly deals with the population group with a large cohort size, including the baby boomer generation born in the late 1940s, in which retirement migration has been particularly noticeable.

Yamada's paper in Chap. 3 is an important work that comprehensively clarifies the residential migration necessitated by the Great Eastern Japan Earthquake in 2011, based on the population migration data published in the 2015 Census. Although most internal migration in contemporary Japan has been voluntary, it is significant in that it is a typical form of displacement migration. The 2015 Census was originally planned to exclude the question on the respondent's address five years ago, which serves as the source of population migration data. However, in order to learn more about the damage caused by the earthquake, it was included among the survey items at the last minute. The author analyzed these data in detail and presented the results in several detailed maps, which is an important feature of the paper. The author argues that residential mobility in the affected areas can be viewed as a process carried out by migrants to recover their daily lives, and the different methods used to secure "safety" for each victim make this mobility complicated and diverse.

The paper of Kanda et al. in Chap. 4 reports on the transitions in the urban structures of 109 cities in Japan based on the constituent municipal population of their functional urban regions from 1980 to 2015. Since the 1980s, Japan's population has become more concentrated in the Tokyo metropolitan area, leading to a wider disparity with other regions. How to remedy this problem of mono-polar concentration in Tokyo has been an important concern in domestic regional policies over the last three decades. The authors examine how Japanese metropolitan areas have undergone structural changes in terms of whether the spatial-cycle model and the differential urbanization theory proposed based on Western experience apply to Japan. In conclusion, they claim that the disurbanization phase of the spatial-cycle model is not observed. Population decline occurred earlier in smaller cities than in the larger cities, reflecting the changes in urban structure, in contrast to the prediction of differential urbanization theory. This finding demonstrates the uniqueness of Japan's experience in relation to those of Europe and the USA.

Sakuno's paper in Chap. 5 discusses the significance and possibility of "relationship population," a new population concept developed in Japan. In contemporary Japan, the majority of municipalities outside of the Tokyo area are facing severe population decline, and there is a strong sense of exasperation in the regions. To overcome this situation and revitalize the regions, we need to view the population of a specific municipality not from the perspective of the resident population, as in the past, but from the relationship population. This approach in Japan of viewing a population in terms of the diverse relationships individuals have established with municipalities and regions other than their primary home has attracted wide attention. The author is a leading geographer conducting research on municipalities and communities in rural areas, where population decline and aging are progressing. On the basis of his previous works, the author examines in detail the various possible interpretations of the framework of this new population concept. It can be said that this paper was the result of efforts to comprehend the current situation and explore a future vision for local governments in rural areas where population decline is becoming a serious problem.

The second volume focuses on Japan's minority populations, including foreign residents and LGBT persons, as well as future visions for Japan and for specific regions and the important politico-economic perspective for studying the current/future Japanese population.

The paper of Takeshita et al. in Chap. 1 discusses the residential patterns of Turkish residents living in Aichi Prefecture. In Japan, interest in international migration and foreign residents has increased along with the progress of population decline. In particular, the formation of ethnic enclaves has attracted wide attention as a theme for geographic studies. The three authors have studied ethnic enclaves using a combination of population census microdata and participant observations. Enclaves of nationalities with a population of more than 100,000 are already well understood. Unfortunately, there has been little prior study on the formation of enclaves of nationalities with smaller populations. This chapter examines dispersed residential patterns from the standpoint of heterolocalism, using the case of Turkish residents. The authors believe that such a point of view may also be applicable to many other small nationality groups in Japan.

Yamauchi's paper in Chap. 2, in contrast to the previous chapter, is the result of studies on sexual minority populations, which have not been studied extensively by population geographers in Japan. The author, who has been studying the population geography of birth for many years, is the first to work on this minority population group. There is a certain volume of research accumulated in Europe and the USA in regard to LBGT persons, where they have been reported to have formed enclaves in large cities. This paper, which is a pioneering article on the population geography of LGBT persons in Japan, examines whether such enclaves can also be found in Japan, particularly in Osaka City. The author concludes that there is no significant association between the presence of LGBTs among respondents and distinct areas where LGBTs are found to be concentrated. Although it is not easy to clarify the geographic distribution of population groups with few members, it is noteworthy that this paper carefully elucidates their distribution by combining a few statistical methods. This is an analytical method that can be applied to studies of other minority groups in similar situations and provides many insights into them.

Koike's paper in Chap. 3 discusses the future of population distribution in Japan using data from regional population projections. In Japan, since the 1980s, the mono-polar concentration into Tokyo has continued, widening the disparity between the Tokyo area and the other regions. This has attracted much attention as one of the biggest problems in Japan, and considerable policy actions have been made to rectify it. It is important to note that the study is based on population projections of geographic units such as prefectures and municipalities conducted by IPSS, and data of projections up to 2045 have already been released. Furthermore, the author has played a central role in making the projections, and in this chapter, he uses these projection data and other sources to examine in detail the possible future population distribution. Interestingly, although he takes into account the increasing population trend of foreign residents in peripheral areas, the author concludes that the mono-polar concentration into Tokyo will continue in the future.

While Chap. 3 examines Japan's population distribution up to 2045 for the entire country, Tanimoto's paper in Chap. 4 projects accessibility to hospital beds up to 2025 for the northern part of the Osaka metropolitan area. IPSS has made projections for each municipality using only the single variable of population. Meanwhile, this paper is exemplary in that it has intentionally carried out regional projections in geographic units below the municipality level. It should be noted that such a detailed study was made possible by the work of Takashi Inoue, the author of Chap. 1 of Volume 1, on population projections at the small area (*chocho-aza*) level covering the entire country. In Japan, which is now in an era of depopulation, there is growing interest in proximity to various social services, including medical care at the community level, below the municipality level. This chapter is of great interest as a pioneering study on an increasingly important topic in Japanese population geography.

The next chapter, Chap. 5, draws attention to the importance of a politico-economic perspective in Japanese population geography. The author of this chapter has published many excellent papers on migration, focusing on life course at the individual level. The paper explores the possibility of establishing politico-economic population geographies by examining how actors in various regions present the problems associated with population decline, on the basis of "The Shock of a Shrinking Japan" published in 2017. In the past, Japanese geographers have been nearly oblivious to the political position of their research. In recent years, however, it has become necessary to discuss policy implications in the publication of results of population geographic studies. Nakazawa's paper is significant in that it compensates for the weaknesses of Japan's traditional population geographic studies and clarifies directions for the future by developing a more in-depth critical examination and discussing the importance of geopolitics in population research.

It is my sincere hope that this anthology of recent outstanding works in Japanese population geography will be read with interest by many people all over the world. Finally, we'd like to express our heartfelt thanks to the Japan Society for the Promotion of Science (JSPS), which made the research leading to the publication of this book possible (No. 21H00637). We are also indebted to Ron Read of the Osaka Branch of Human Global Communications Co., Ltd., for carefully and kindly editing the earlier English manuscripts.

Kyoto, Japan   Yoshitaka Ishikawa

Contents

Chapter 1
Internal Migration in the Post-Demographic Transition Period

Takashi Inoue

Abstract This chapter analyzes the internal migration in post-war Japan and exam-ines the influence of the demographic transition on so-called migration turnarounds in the post-demographic transition period. We grasp migration turnarounds and their factors chiefly by observing the long-term change of mainstream migration and coun-terstream migration. These two kinds of migration are defined respectively as migra-tion from non-metropolitan areas to metropolitan areas and from metropolitan areas to non-metropolitan areas. The analyses are performed using three demographic indi-cators, i.e., Migration Effectiveness Index, Migration Preference Index, and Cohort Cumulative Social Increase Ratio, whose fundamental concept was proposed by the author. Consequently, the following conclusions are reached: (1) The rapid decrease in fertility in the final phase of the demographic transition caused a reduction of main-stream migration after about 20 years and a reduction of counterstream migration after about 25 years. As a result, this led to the two obvious migration turnarounds. (2) Other migration turnarounds, such as the population concentration in the Tokyo metropolitan area and the strong decline in return migration, were directly caused by economic factors. These factors can be partly explained by the demographic bonus and onus, which were caused by the demographic transition and the persistent decrease in fertility in the post-demographic transition period.

Keywords Internal migration · Demographic transition · Migration turnaround · Post-demographic transition period · Mainstream migration · Counterstream migration

T. Inoue (✉)
Department of Public and Regional Economics, College of Economics, Aoyama Gakuin University, Shibuya-ku, Tokyo 150-8366, Japan
e-mail: t-inoue@cc.aoyama.ac.jp

© The Author(s), under exclusive license to Springer Nature Singapore Pte Ltd. 2023
Y. Ishikawa (ed.), *Japanese Population Geographies I*,
Population Studies of Japan,
https://doi.org/10.1007/978-981-99-2035-8_1

1

1.1 Introduction

This chapter analyzes Japan's postwar internal migration from various perspectives and discusses how the country's demographic transitions—in particular, the declining birth rate and mortality rate—affected the so-called migration turnaround during the post-demographic transition period. According to Ishikawa (2001), a migration turnaround is a major change in the nationwide spatial pattern of migration, specifically, the emergence of a pattern contrary to conventional expectations. In other words, Japan's internal migration trends are reversing. Typical examples include counterurbanization,[1] which is a counterstream that occurs after the phenomenon of dominant rural-to-urban migration (i.e., urbanization), and reurbanization, which occurs after counterurbanization. As discussed below, at least two such migration turnarounds have been observed in Japan since the 1960s. However, although many studies have dealt with trends in internal migration in post-war Japan, not so many have dealt with such migration turnarounds.[2]

Meanwhile, Wilber Zelinsky's hypothesis of mobility transition is a well-known concept that is similar to migration turnaround (Zelinsky 1971). This hypothesis posits that, in the process of social modernization, the quantities and patterns of migration will also change, just as seen by the demographic transitions caused by declining birth and death rates. However, in this context, mobility transitions refer to migration over extremely long periods of time, extending from pre-modern society to mature societies of the future,[3] which means this is a more abstract and broader concept than the migration turnaround described by Ishikawa (2001). In addition, these hypotheses have been presented without any rigorous verification regarding relevance to demographic transitions. In contrast, this chapter limits the period under discussion to the post-demographic transition period to empirically examine the relationship between demographic transitions and internal migration. Therefore, the approach here applies migration turnaround defined by Ishikawa (2001) rather than the mobility transition defined by Zelinsky (1971).

Section 1.2 of this chapter discusses the characteristics and causes of internal migration in postwar Japan based on data for migration between municipalities from the publication "Annual report on the internal migration in Japan derived from the basic resident registers" issued by the Statistics Bureau of the Ministry of Internal Affairs and Communications. Section 1.3, using this report's data on inter-prefectural migration, ascertains the characteristics of internal migration in post-war Japan by calculating the Migration Effectiveness Index and Migration Preference Index, which are representative indicators of migration. Then, Sect. 1.4 analyzes the Cohort Cumulative Social Increase Ratio (CoCSIR) using the age-specific census population together with the publication "Regional Population Projections for Japan (March 2013 estimate)" from the National Institute of Population and Social Security Research. Finally, the results are summarized by referring to the relationship between migration turnaround and demographic transition based on the discussions in Sects. 1.2–1.4.

1.2   Characteristics and Causes of Internal Migration in Post-War Japan

After grasping the long-term changes in the amount and rate of migration and the sex ratio of migrants in parts 1 and 2 of this section, part 3 discusses migration turnaround and its causes by focusing on the changes in mainstream migration and counterstream migration.

(1)   Long-term trends in the quantity and rate of migration

Internal migration in Japan accelerated rapidly from the 1960s to the early 1970s, and both the number of migrants (about 8.54 million in 1973) and the migration rate (about 8.02% in 1970) peaked at the end of this period. Thereafter, although a weak peak was observed around 1995, there was a gradual decline that basically continued until the 2010s, and by 2013, the number of migrants had fallen to about 5.02 million, and the migration rate had fallen to about 3.99%. When the economy grows in a virtuous cycle, employment opportunities increase, and thus migration inevitably becomes active. Therefore, the increased mobility of internal migration, which peaked in the early 1970s, can be explained by rising employment during Japan's period of high economic growth (December 1954 to November 1973). On the other hand, during the bubble economy period (December 1986 to February 1991), there was a similar economic boom but no increase in mobility whatsoever. The reasons for this are discussed in Sect. 1.3.

When internal migration is classified into intra-prefectural and inter-prefectural migration, these two migration rates were nearly equal during the period of rapid economic growth from the 1960s to the early 1970s, whereas in other times, the intra-prefectural migration rate was higher. This means that inter-prefectural (i.e., long-distance) migration was prevalent during the period of rapid economic growth, except in the initial stages. This is consistent with the large-scale migration of mainly junior high school graduates from non-metropolitan areas to metropolitan areas through group employment during this period. On the other hand, around 1995, although there was a weak peak in intra-prefectural migration, no such tendency was observed in inter-prefectural migration. In other words, the abovementioned weak peak around 1995 was brought about by an increase in the mobility of intra-prefectural migration.

(2)   Long-term trends in the sex ratio of migrants

Figure 1.1 shows how the sex ratio of migrants varies in intra- and inter-prefectural migration. The sex ratio of migrants is an index expressed as {the number of male migrants} ÷ {the number of female migrants} × 100. A sex ratio greater than 100 thus indicates that there are more male migrants than female migrants.

All of the values in this figure are above 100 except for the intra-prefectural migration in 1959, indicating that the mobility of internal migration in Japan is basically higher for men than for women. Next, a comparison of intra-prefectural and inter-prefectural migration shows that the sex ratio of the former is much lower than that of the latter and is close to 100. This is because intra-prefectural migration

Fig. 1.1 Sex ratio of migrants in intra- and inter-prefectural migration (1958–2012)

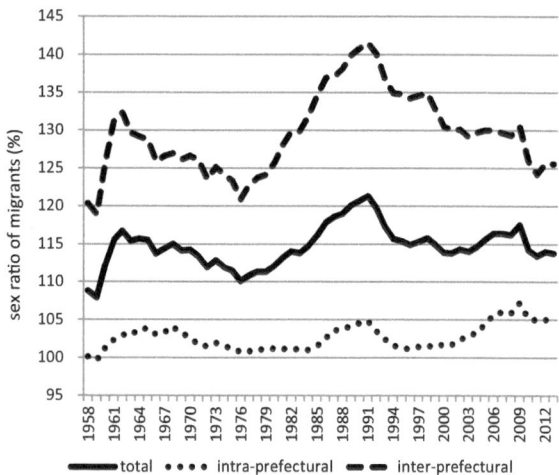

generally involves a higher percentage of household migration due to house purchases or changes of residence without an associated change in the social status of the migrant, while inter-prefectural migration generally involves a higher percentage of individual migration due to such life events as starting higher education/employment or transferring to a new job or company department, which generally involves a change in the social status of the migrant. It has been pointed out that household migration naturally involves the movement of men and women in equal numbers, while it is mainly men who travel long distances for higher education and employment (Inoue 2001).

Furthermore, changes in the sex ratio of migrants show that both intra-prefectural and inter-prefectural migration increased significantly during both the period of high economic growth and the bubble period. This shows that the mobility of male workers in particular was much larger due to the expansion of employment accompanying economic growth. It should also be noted that although the scale of the economy was greater during the period of high economic growth compared with the bubble period, the sex ratio of migrants was larger in the latter case. The reason for this phenomenon can be explained as follows: While heavy industry, which was dominated by male workers, grew remarkably during the period of rapid economic growth, there was also significant growth in light industry (i.e., the textile industry), which required large numbers of female workers; however, during the bubble period it appears that the major growth areas were industries such as finance and real estate, which were dominated by male workers.

(3) **Long-term transitions and migration turnaround of mainstream and counterstream migration**

In general, mainstream migration refers to rural-to-urban migration, while counterstream migration refers to urban-to-rural migration. Here, by observing the long-term transitions of mainstream and counterstream migration, we consider the

Fig. 1.2 Long-term trends in four types of migration (1954–2012)

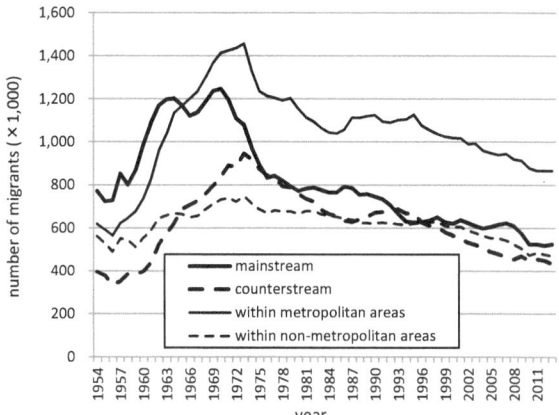

migration turnaround and its causes with regard to internal migration in Japan. Figure 1.2 shows the long-term trends in four types of inter-prefectural migration: non-metropolitan to metropolitan migration, metropolitan to non-metropolitan migration, intra-metropolitan migration, and intra-non-metropolitan migration. Here, the term "metropolitan" refers to all eleven prefectures that make up the Tokyo metropolitan area (Saitama, Chiba, Tokyo, Kanagawa), the Nagoya metropolitan area (Gifu, Aichi, Mie), and the Osaka metropolitan area (Kyoto, Osaka, Hyogo, Nara). Migration from non-metropolitan to metropolitan areas corresponds to mainstream migration, while movement from metropolitan areas to non-metropolitan areas corresponds to counterstream migration.

From Fig. 1.2, the following changes can be observed in the relationship between the sizes of mainstream and counterstream migration: (1) from the 1960s to the early 1970s, mainstream migration greatly exceeded counterstream migration; (2) in the late 1970s, mainstream and counterstream migration were balanced due to a sharp decline in mainstream migration; (3) in the 1980s, mainstream migration exceeded counterstream migration due to a decline in counterstream migration; (4) from 1993 to 1995, counterstream migration exceeded mainstream migration; and (5) since 1996, mainstream migration has surpassed counterstream migration. Of the four points in time at which these changes occurred (1975, 1980, 1993, and 1996), there were relatively large trend changes in 1975 and 1980, suggesting that migration turnarounds occurred at least at these two points in time.

In the following, we first discuss the migration turnaround that took place around 1975.[4] During the period of rapid economic growth from the 1960s to the early 1970s, the rapid increase in employment opportunities in industrial metropolitan areas led to a sharp increase in mainstream migration, and the amount of net in-migration to metropolitan areas reached a maximum of 650,000 people per year. From around 1975, however, the mainstream and counterstream started to balance out. This change in the trend is generally attributed to return migration becoming dominant due to the transition from a period of high economic growth to one of low growth, coupled with increasing pollution in metropolitan areas.[5] Despite this apparent conclusion, the change was in fact caused by the drastic decrease in mainstream migration described above. Furthermore, the main cause of this sharp decline in mainstream migration was found to be the cohort effect (Ishikawa 1994), according to which the differences in cohort size give rise to various demographic phenomena. Here, we focus on the size of the population aged 15–19 years in non-metropolitan areas, who are the principal agents of mainstream migration. However, this population group does not include people who have already performed mainstream migration into the metropolitan area. Therefore, when discussing the size of this population, it makes sense to also discuss its size immediately before leaving home and embarking on mainstream migration. In other words, it is necessary to examine the size of the population aged 10–14 years in non-metropolitan areas. In general, although members of the population aged 10–14 years have significantly lower mobility than those in its adjacent older and younger age groups, this age group is expected to soon generate most of the people leaving home for mainstream migration. Therefore, the population aged 10–14 years in non-metropolitan areas becomes the parent body of the population group who later engage in mainstream migration.

Table 1.1 shows the population aged 10–14 years nationwide and in non-metropolitan areas for each cohort starting with the 1936–40 cohort. In the 1960s and early 1970s, people reaching the 15–19 years age group consisted of the 1946–1950 cohort, which includes the baby boomers (born during the first baby boom of 1947–49) and its neighboring cohorts. The people reaching the 15–19 years age group in the late 1970s consisted of the 1956–60 cohort. Therefore, a comparison of the 1946–50 and 1956–60 cohorts based on Table 1.1 shows that the national populations of these cohorts were 11.13 million and 7.978 million, with the non-metropolitan populations accounting for 7.209 million and 4.906 million. This means that the latter cohort is a little over 70% and a little under 70% of the former one in the cases of the national population and the non-metropolitan population, respectively. Meanwhile, the minimum value of mainstream migration in the latter half of the 1970s was about 64% of the maximum value of mainstream migration in the period of high economic growth, indicating that the drastic decrease during this period can be mostly explained by the cohort effect.

Next, we consider the migration turnaround that occurred around 1980. In the 1980s, mainstream migration once again began to outpace counterstream migration due to a phenomenon known as the reconcentration of population in cities, which was observed in other developed countries at about the same time (Ishikawa 2001). However, as mentioned above, this phenomenon was not caused by an increase in mainstream migration but by a decrease in counterstream migration. Furthermore, this phenomenon can also be explained by the cohort effect. The counterstream consistently declined from 1973 to 1987, but it is thought that the cohorts who were in their mid-20s during this period[6]—i.e., the population from the 1946–50 baby boomer cohort to the 1961–65 cohort—actively performed return migration and became the mainstay of the counterstream. According to Table 1.1, there was a downward trend in population size from the 1946–50 cohort to the 1961–65 cohort, which had led to a reduction in the counterstream.

Table 1.1 National and non-metropolitan population aged 10–14 years by cohort

Cohort	Year	National population (×1000)	Non-metropolitan population	
			×1000	Share (%)
1936–40	1950	8812	6003	68.1
1941–45	1955	9585	6232	65.0
1946–50	1960	11,130	7209	64.8
1951–55	1965	9318	6107	65.5
1956–60	1970	7978	4906	61.5
1961–65	1975	8285	4554	55.0
1966–70	1980	8965	4590	51.2
1971–75	1985	10,046	5114	50.9
1976–80	1990	8548	4555	53.3
1981–85	1995	7485	4059	54.2
1986–90	2000	6558	3534	53.9
1991–95	2005	6036	3153	52.2
1996–00	2010	5966	2999	50.3

Source Census
Note 1936–40 cohort means the population group born from October 1935 to September 1940 (similarly for all other cohorts)

This suggests that the two migration turnarounds that occurred in post-war Japan were both caused by differences in cohort size between the baby boomer generation and subsequent generations. According to this analogy, the 1971–75 cohort, which, like the 1946–50 cohort, had a size of over 10 million people (Table 1.1), should also have exhibited increased mainstream migration on reaching the 15–19 years age group. However, around 1990, when the 1971–75 cohort reached the age of 15–19 years, mainstream migration was actually on a downward slide (Fig. 1.2). The reasons for this phenomenon can be explained as follows. The 1971–75 cohort includes junior baby boomers (second baby boomers born in 1971–74), many of whom were born as children of baby boomers who had moved to metropolitan areas. As a result, the population of this cohort living in non-metropolitan areas aged 10–14 years (5.114 million) was only about 70% of the corresponding number for the 1946–50 cohort. As can be seen from Table 1.1, the 1971–75 cohort's share of the non-metropolitan population aged 10–14 was one of the lowest (second only to the 1996–2000 cohort, which was the youngest), suggesting that most of the junior baby boomers were born in metropolitan areas.

Finally, some reference should be made to mobility within metropolitan areas and mobility within non-metropolitan areas. The intra-metropolitan migration shows a pattern similar to that of counterstream migration, peaking in 1973 and continuing a gradual downward trend thereafter. The appearance of this migration peak immediately after the end of the high economic growth period is thought to be due to the addition of return migration within metropolitan areas in addition to migration to the suburbs by people who had migrated into metropolitan areas. For example, the ratio of the amount of migration from Tokyo to its three surrounding prefectures to the number of intra-metropolitan transfers increased from about 20% in 1956, when the number of intra-metropolitan transfers was the lowest, to about 31% in 1973. On the other hand, although non-metropolitan migration peaked in 1973 and has been slowly declining since then, the overall change in this type of migration is smaller than in other types of migration and would be less affected by economic cycles.

1.3   Analysis Based on Migration Effectiveness Index and Migration Preference Index

In this section, we use the Migration Effectiveness Index and Migration Preference Index, which are representative indicators of population migration, to understand the characteristics of internal migration in post-war Japan.

(1)   Analysis based on Migration Effectiveness Index

The Migration Effectiveness Index expresses the ratio of migrants who have changed the population distribution of a region to the total number of inter-regional migrations during a given time period. If I_i is the total amount of in-migration into region i, and O_i is the total amount of out-migration from region i, the Migration Effectiveness Index E is expressed by the following formula (multiply by 100 to express it as a percentage):

$$E = \frac{\sum_i |I_i - O_i|}{\sum_i (I_i + O_i)} = \frac{\sum_i |I_i - O_i|}{2 \sum_i I_i} = \frac{\sum_i |I_i - O_i|}{2 \sum_i O_i}$$

The Migration Effectiveness Index does not necessarily correlate to the mobility level, i.e., the migration rate. Accordingly, if there are nearly equal amounts of in-migration and out-migration, the Migration Effectiveness Index approaches zero. Applying this index to migration between metropolitan and non-metropolitan areas, we obtain Migration Effectiveness Index = |the amount of mainstream migration —the amount of counterstream migration| ÷ (the amount of mainstream migration + the amount of counterstream migration).

Figure 1.3 shows the long-term changes in the Migration Effectiveness Index (%) from 1954 to 2013 for inter-prefectural migration and migration between metropolitan and non-metropolitan areas. Obviously, the latter type of migration partially coincides with the former type in cases of inter-prefectural migration where people cross the boundaries between metropolitan and non-metropolitan areas. According to this figure, both indices are well linked, although the index for the latter has a larger range of change. Changes in the index for inter-prefectural migration can be largely explained by changes in migration between metropolitan and non-metropolitan areas. The changes in these indices show three distinct peaks in the late 1950s/early 1960s, the late 1980s, and around 2002–2008. These peaks occurred

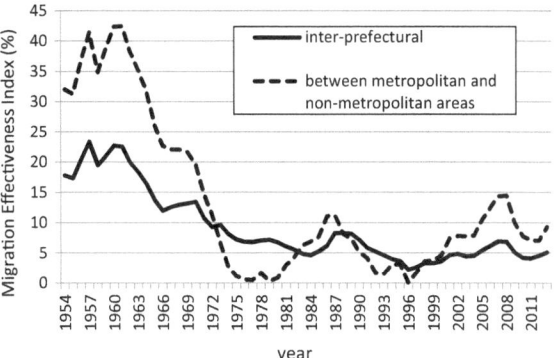

Fig. 1.3 Long-term trends of Migration Effectiveness Index (1954–2012)

during the high economic growth period, the bubble period, and the Izanami Boom period (February 2002 to February 2008), respectively, suggesting that the index is boosted by virtuous economic cycles. This is attributed to the boom times stimulating migration to areas where growth industries are currently located.

Next, let us confirm the extent to which the Migration Effectiveness Index is linked to the migration rate. In the period of high economic growth, the index declined sharply from the middle to the latter half of the growth period (1960s to early 1970s), whereas the migration rate showed the opposite trend and increased during this period. This contrasting behavior is thought to be due to a rapid increase in the counterstream during the same period and the narrowing of the gap with the main-stream (Fig. 1.2). Moreover, the inter-prefectural migration rate either changed little or declined during the bubble period and the Izanami Boom period. This phenomenon is attributed to the fact that a booming economy has the effect of persuading people who moved to metropolitan areas to remain in these areas. While this effect reduces the migration rate, it also reduces the counterstream, widening the gap between the mainstream and the counterstream and pushing up the index.

(2) Analysis based on Migration Preference Index

The Migration Preference Index is the ratio of the actual number of migrations to the expected number of migrations calculated based on the assumption that the number of migrations is proportional to the population size of the origin and destination. If PI_{ij} is the Migration Preference Index from region i to region j, then PI_{ij} is expressed by the following equation:

$$PI_{ij} = \frac{M_{ij}}{\frac{P_i}{P_T} \cdot \frac{P_j}{P_T - P_i} \cdot \sum M_{ij}} \times 100$$

Here, M_{ij} is the number of migrants from region i to region j, $\sum M_{ij}$ is the total number of migrants in the entire target region, P_i and P_j are the populations of regions i and j, respectively, and P_T is the total population of the target region. If i and j are prefectures in Japan, then P_T is the total population of Japan. If this value exceeds

100, the actual number of migrants is more than the expected number, indicating a strong connection between areas i and j. In addition, although it is generally known that a shorter distance between the origin and destination corresponds to a greater number of migrants, the Migration Preference Index does not include information on distance. Therefore, the value of this index is expected to be larger if the distance between regions i and j is shorter.

The results of calculating the Migration Preference Index for inter-prefectural migration from all over Japan to Tokyo for the three years of 1970, 1990, and 2010 are shown in Fig. 1.4. These values ought to show the extent to which people prefer Tokyo as a migration destination, or in other words, the strength of their preference for Tokyo. According to this figure, regional differences in the Migration Preference Index exhibit the following four characteristics: (1) The values are generally higher in eastern Japan. (2) This is particularly so in the three prefectures surrounding Tokyo (Saitama, Chiba, Kanagawa). (3) The values are also relatively high in Miyazaki, Kagoshima, and Okinawa, which are the farthest from Tokyo. (4) The values are low in other prefectures including Fukui, Gifu, Mie, Shiga, Nara, Wakayama, and Tokushima. Of these, the prefectures in (1) are generally closer to Tokyo than those of western Japan, and the prefectures in (2) are adjacent to Tokyo. As for (3), taking Okinawa as an example, although it is a long way from Tokyo, it is also a long way from other attracting cities such as Osaka and Fukuoka, and thus people often choose Tokyo for its larger urban scale. As for (4), Gifu and Mie prefectures are close to Nagoya, while Fukui, Shiga, Nara, Wakayama, and Tokushima prefectures are located close to major cities in the Osaka area, which people often choose as destinations.

Next, let us consider annual changes in this index. According to Fig. 1.4, in the Hokkaido, Tohoku, North Kanto (Ibaraki, Tochigi, Gunma), Koshinetsu (Niigata, Yamanashi, Nagano), and Kyushu (except Fukuoka) regions, the index tended to decline over this 40-year period, whereas in places such as the Nagoya metropolitan area, the Kinki region (including the Osaka metropolitan area), Okayama, Hiroshima, and Fukuoka, the index generally tended to increase. These trends can be interpreted as follows. During the period of rapid economic growth, the major cities in western Japan had sufficient population attraction, so there was a strong preference for Tokyo in eastern Japan and the Kyushu region, which were less influenced by these cities. After that, since the population attraction of cities in western Japan became relatively weak, it seems that the attraction of Tokyo became stronger in metropolitan areas around these cities. In 1970, at the end of the period of rapid economic growth, the Tokyo metropolitan area accounted for about 60% of net in-migration to metropolitan areas. In both 1990 and 2010, however, the percentage rose to over 85%. Thus, over the past 40 years, a kind of migration turnaround has been going on, resulting in the population concentration in Tokyo. This suggests that it is more reasonable to focus on migration between the Tokyo metropolitan area and all the other areas rather than that between metropolitan and non-metropolitan areas in order to grasp the internal migration trends in Japan since the period of rapid economic growth.[7]

Fig. 1.4 Migration
Preference Index values for
migration into Tokyo from
other prefectures in 1970,
1990, and 2010

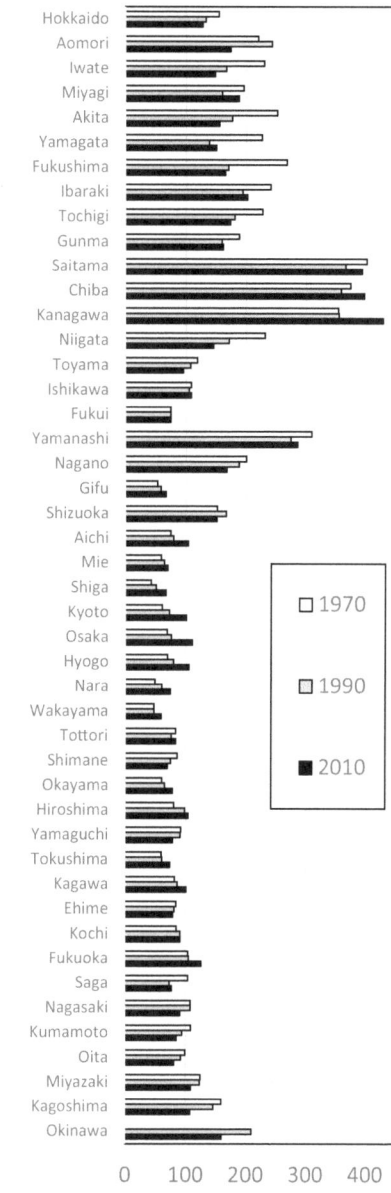

1.4 Analysis Based on Cohort Cumulative Social Increase Ratio (CoCSIR)

This section uses the Cohort Cumulative Social Increase Ratio (CoCSIR), the basic concept of which was presented by Inoue (2002), to find the characteristics of internal migration during the post-demographic transition period.

(1) What is CoCSIR?

The CoCSIR is an index defined as the ratio of the cumulative number of social increases in a cohort since age 10–14 years to the population of this cohort when aged 10–14 years.[8] In the formula of the CoCSIR, the population of the cohort when aged 10–14 years is used as the denominator and is regarded as the reference population for the cohort. This is because, as mentioned above, the 10–14 years age group has the following two characteristics: one is that this group has significantly lower mobility than the preceding and following age groups; the other is that, in the preceding age group (15–19 years age group), the number of people leaving home rapidly increases and thus the 10–14 years age group can be regarded as the population chiefly responsible for this sort of migration. In addition, the 10–14 years age group includes junior high school students, and studies of return migration often assume that most Japanese people consider their hometown to be the place where they lived while attending junior high school.[9] Therefore, this chapter considers the 10–14 years age group living at a given place to represent the original population of the corresponding cohort at that place. Based on this assumption, the CoCSIR can be regarded as the ratio of the cumulative total of subsequent social increases relative to the original population.

In general, most people who move between metropolitan and non-metropolitan areas in Japan (both the mainstream and counterstream) are from non-metropolitan areas (Inoue 2002). Therefore, if all migrants that move between metropolitan and non-metropolitan areas were considered to originate from the non-metropolitan areas, then the CoCSIR for non-metropolitan areas could be used to track long-term out-migration rates, since it indicates how many people from non-metropolitan areas have moved elsewhere (i.e., are staying in metropolitan areas) at any given time. This indicator converges to a certain value as people age, and the value for the non-metropolitan areas at that time can be regarded as the lifetime exodus rate for people from non-metropolitan areas.[10] Empirical studies using CoCSIR, such as Shimizu (2006, 2009), have also confirmed the usefulness of this approach. Inoue (2014) attempted to formulate this indicator and to explain its theoretical background.

The cumulative number of social increases that forms the numerator of this index can be calculated by the life table survival rate method or the inter-census survival rate method. In particular, the latter method can be calculated from the population in each age group according to the census. Since this is a simple method, it is also used in this chapter.[11]

(2) Characteristics of internal migration by cohort

Here, after calculating the CoCSIR for twelve cohorts from the 1941–45 cohort to the 1996–2000 cohort for metropolitan and non-metropolitan areas,[12] we use the results to understand the characteristics of internal migration. Figure 1.5 shows the calculation results. For the sake of clarity, the results are spread across four figures (a)–(d) showing three cohorts each. In this figure, the horizontal axis represents the age group reached by the cohort, and the vertical axis represents the CoCSIR value at that time. For example, the labels *m1941–45* and *n1941–45* in this figure refer to values for the 1941–45 cohort in the metropolitan and non-metropolitan areas, respectively. Thus, for example, the data for *m1956–60* at age 25–29 years represents the cumulative total social increase in the 1956–60 cohort in metropolitan areas from the 10–14 years age group until it reached the 25–29 years age group (or in other words, the cumulative value of the number of social increases from 1970 to 1985) divided by the population of this cohort when aged 10–14 years. Naturally, this value is always positive for metropolitan areas and negative for non-metropolitan areas, and thus in each figure the upper three lines correspond to metropolitan areas and the lower three lines correspond to non-metropolitan areas. In addition, since the cumulative social growth of metropolitan areas is exactly opposite in sign to the cumulative social growth in non-metropolitan areas, the two values change roughly symmetrically.[13]

According to Fig. 1.5, apart from the change from the 1971–75 cohort to the 1981–85 cohort (Fig. 1.5c), the values tend to approach zero for the later cohorts, indicating a significant decline in mobility related to travel between metropolitan and non-metropolitan areas. This means that in later cohorts, there is a lower percentage of people migrating from non-metropolitan areas to metropolitan areas. Meanwhile, when observing how the values change with aging in metropolitan areas for a fixed cohort, the cohorts prior to the 1976–80 cohort show a rapid increase from the 15–19 years age group to the 20–24 years age group and then immediately decline between the 20–24 and 25–29 years age groups, whereas the 1981–85 and later cohorts do not show a clear downward trend after the age of 20–24 years. For migration between metropolitan and non-metropolitan areas in all age groups, almost no net out-migration was observed in metropolitan areas and almost no net in-migration was observed in non-metropolitan areas. This implies that a kind of migration turnaround occurred.

In the following, we consider the factors that caused this migration turnaround. Needless to say, the reduction of these values in the 20–24- and 25–29-year age ranges in cohorts prior to 1976–80 is chiefly owing to return migration by non-metropolitan residents who had been staying in metropolitan areas. Therefore, it is not possible to explain this migration turnaround unless it is assumed that return migration decreased significantly for cohorts after the 1981–85 cohort.[14] Since the 20–24 and 25–29 years age groups were reached during 2000–2005 by the 1976–80 cohort and during 2005–10 by the 1981–85 cohort, it appears that the return migration of people from metropolitan areas greatly declined during the 2000–10 period. Over the past decade, due to progress in the redevelopment of central metropolitan areas, falling land prices, and the development of public transportation systems, the attraction of living in urban cores has risen compared to not only non-metropolitan areas

Fig. 1.5 Change of CoCSIR
in metropolitan and
non-metropolitan areas with
aging

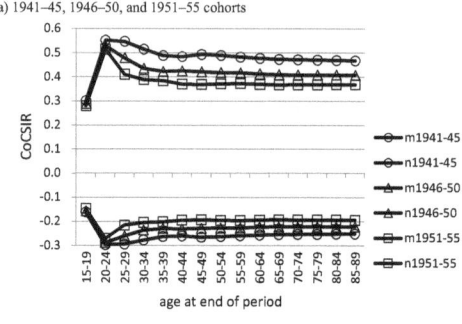

a) 1941–45, 1946–50, and 1951–55 cohorts

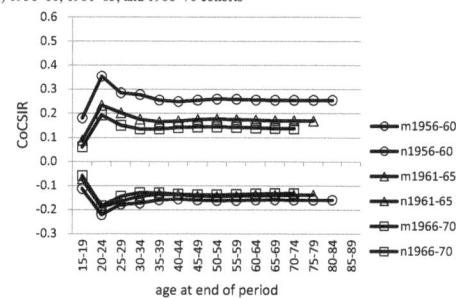

b) 1956–60, 1961–65, and 1966–70 cohorts

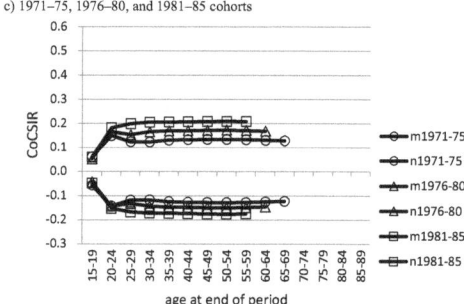

c) 1971–75, 1976–80, and 1981–85 cohorts

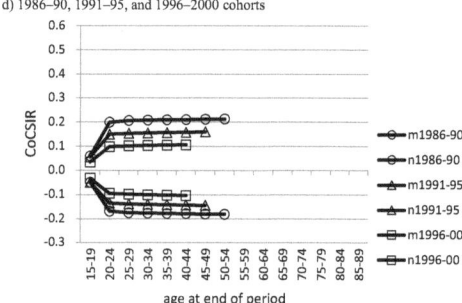

d) 1986–90, 1991–95, and 1996–2000 cohorts

but also the suburbs of metropolitan areas, and it is well known that this phenomenon prompted the trend of returning to urban cores. Ezaki (2006) showed that one of the main causes of this tendency to return to urban cores is the decline in the migration from urban cores to the suburbs, calling this phenomenon the "end of suburban-ization."[15] Of the migrations that originate from the center of metropolitan areas, the migration where the destination is in the suburbs generates suburbanization, and most of the migrations where the destination is in a person's hometown are return migrations. Therefore, it is natural to assume that return migration is also strongly suppressed by the possible factors behind the end of suburbanization. It is thus highly likely that the end of suburbanization and the decline of return migration occurred almost simultaneously in the early twenty-first century.

Finally, using CoCSIR, let us explain the issue raised in Sect. 1.2, i.e., why mobility levels did not increase during the bubble period. One possible reason for this is that the booming economy had the effect of curbing outflows from metropolitan areas. This can be inferred from the following facts: In metropolitan areas during 1985–90, the values for the 1951–55 cohort (equivalent to ages 30–34 to 35–39 years) changed from 0.389 to 0.383, and the values for the 1956–60 cohort (equivalent to ages 25–29 to 30–34) changed from 0.285 to 0.277; these changes were small compared to the previous and subsequent cohorts, and as a result, the metropolitan areas experienced only a slight amount of net out-migration.

1.5   Conclusion

This chapter summarizes how the migration turnaround discussed so far is related to the so-called demographic transition. Japan's demographic transition, of which the final phase was the declining birth rate, is known to have occurred over a very short period of time. Specifically, the TFR plummeted from 4.32 to 2.04 over the period of just over eight years from 1949 to 1957 after the end of the first baby boom, and there is no denying that this period is either the final phase of the demographic transition or at least a part of it.[16] As noted earlier, this decline in fertility is reflected in the difference in population size of those aged 10–14 years between the 1946–50 cohort and the 1956–60 cohort (Table 1.1).

Since the mainstream generally declines sharply after the age of 20–24 years, it is thought that by 1970 and 1980, when both cohorts reached this age, these cohorts mostly exited from the mainstream. Theoretically, therefore, the decline of the mainstream should take place over the decade from 1970 to 1980, between the end of the exit of the larger 1946–50 cohort and the end of the exit of the smaller 1956–60 cohort. In fact, the mainstream did actually decrease during the 1970–80 period, which perfectly matches the theoretical figure.

A similar argument can be made for the reduction of the counterstream. In general, since the counterstream declined sharply after age 25–29 years, by 1975 and 1985, when these two cohorts reached age 25–29 years, they mostly exited from the coun-terstream. As a result, the decade or so between 1975 and 1985 can be regarded as the period of counterstream decline. In fact, the actual period of counterstream decline was from 1973 to 1987, which is slightly longer than the theoretical period but still a close match.

Based on the above discussion, the sharp decline in the fertility rate marking the end of the demographic transition inevitably led to a decline in the mainstream about 20 years later, which was followed by a decline in the counterstream about 25 years after that. Consequently, this development is thought to have resulted in the two migration turnarounds discussed in Sect. 1.2.

On the other hand, the types of migration turnaround pointed out in Sects. 1.3 and 1.4 are thought to be largely due to direct economic factors. The overconcentration of the population in Tokyo, which was pointed out in Sect. 1.3, was largely caused by regional economic disparities between metropolitan areas, while the decline in return migration pointed out in Sect. 1.4 is thought to have been influenced by the reduction in the cost of living in the centers of metropolitan areas. From a broader perspective, however, it is undeniable that these economic factors are driven by the population bonus and onus resulting from the demographic transition and subsequent further decline in the birth rate. In other words, the migration during the post-demographic transition period has been to a greater or lesser extent influenced by the demographic transition, both in terms of quantitative and qualitative changes.

Notes

1. Counterurbanization is the predominance of migration from urban to rural areas, mainly due to the return migration of people who had moved to cities from rural areas. Note that this is a different concept from the so-called suburbanization, where the population concentrated in a city expands into the suburbs.
2. In addition to Ishikawa (2001), see, e.g., Abe (1994), Inoue (2002), and Shimizu (2011).
3. Zelinsky (1971) divided societies into five stages according to their state of migration transition: (1) pre-modern society, (2) early transitional society (dominated by rural-to-urban migration), (3) late transitional society, (4) advanced society (decreased rural-to-urban migration, dominant inter- and intra-urban migration), and (5) future super-advanced society.
4. Although there are various theories regarding the timing of the population shift, this chapter is purely concerned with the reversal of the magnitude relationship between the mainstream and the counterstream. Oe (1995) clearly organized the discussion on this subject.
5. For example, the Japan Labor Research Institute (1994) made these observations.
6. Patterns of age-specific migration rates reveal that outflows in metropolitan areas usually peak in the migrants' mid-20s (Inoue 1991).
7. This chapter focuses on migration between metropolitan and non-metropolitan areas because it was necessary to discuss the issue from a long-term perspective, including the period of rapid economic growth. Although a discussion on the situation after the period of high economic growth should also mention migration between the Tokyo metropolitan area and all the other areas, this has been omitted due to lack of space. See Nakagawa (2005), Shimizu (2011), and Hirai (2014) for more information on the characteristics of this sort of migration.
8. Although the "cumulative net migration rate" proposed by Kawabe (1985) is a similar indicator, this value overestimates the value of the migration rate with a small denominator because the size of the denominator changes with aging. Refer to Hiroshima (2014) for a detailed discussion of these issues, including a comparison with CoCSIR. While CoCSIR was originally called

the cohort cumulative net migration ratio, since this index is not obtained by directly observing migration phenomena, its name was changed to social increase. However, this name change does not change the definition or significance of the indicator, since the net migration numbers and social increase are fully consistent.

9. See, e.g., Esaki et al. (1999, 2000).

10. This value indicates the rate of people from non-metropolitan areas who eventually settled in a metropolitan area. To be precise, it corresponds to the value obtained by taking the minus sign away from the value for non-metropolitan areas. According to the hypothesis on people potentially expected to leave their hometowns presented by Ito (1984), "average number of siblings minus 2" would define this rate. This hypothesis is discussed in detail by Maruyama and Oe (2008) and Nakagawa (2010).

11. Since it is necessary to calculate this indicator for as many cohorts as possible, this chapter used the National Institute of Population and Social Security Research's forecasted population by prefecture and age group in addition to the National Population Census, as described in the introduction. There are three types of inter-census survival rate methods: the forward method, the backward method, and the average method (Yamaguchi 1989). This chapter used the forward method.

12. The 1941–45 cohort refers to the population group born between October 1940 and September 1945. The same applies to all of the other cohorts.

13. Since the denominators are different, the form is not perfectly symmetrical.

14. Here, return migration refers to cases where someone has lived for no longer than ten years in the metropolitan area. However, it has also been pointed out that return migration by elderly people who have lived in cities for much longer is expected to increase (Hirai 2011).

15. According to Ezaki, this phenomenon has definitely become more pronounced since the late 1990s.

16. This is discussed in detail by, for example, Ato (2000).

References[1]

Abe T (1994) Kokunai jinkoido ni okeru shuryu to gyakuryu no taimu ragu (The time lag between dominant streams and reverse streams within internal migration in Japan). Jinkogaku Kenkyu (The Journal of Population Studies) 17:33–40 https://doi.org/10.24454/jps.17.0_33 (J)

Atoh M (2000) Gendai jinkogaku: Shoshi korei shakai no kiso chishiki (Modern demography: Basic knowledge of the aging society with low fertility). Nihon Hyoronsha, Tokyo (J)

Esaki Y (2006) Shutoken jinko no shoraizo: Toshin to kogai no jinko chirigaku (The prospect of Greater Tokyo's population: Population geography of urban core and suburbs). Senshu University Press, Tokyo (J)

Esaki Y, Arai Y, Kawaguchi T (1999) Jinko kanryu gensho no jittai to sono yoin: Nagano ken shusshin dansei o rei ni (Return migration from major metropolitan areas to Nagano Prefecture). Chirigaku Hyoron (Geographical Review of Japan) 72A(10):645–667 https://doi.org/10.4157/grj1984a.72.10_645 (J)

Esaki Y, Arai Y, Kawaguchi T (2000) Chihoken shussinsha no kanryu ido: Nagano ken oyobi Miyazaki ken shusshinsha no jirei (Return migration in Japan: A comparative analysis of migrants returned to Nagano and Miyazaki Prefectures). Jimbun Chiri (Japanese Journal of Human Geography) 52(2):190–203 https://doi.org/10.4200/jjhg1948.52.190 (J)

Hirai M (2011) Korei jinko no bunpu to ido (Distribution and migration of elderly population). In: Ishikawa Y, Inoue T, Tahara Y (eds) Chiiki to jinko kara miru Nihon no sugata (The population geography of contemporary Japan). Kokon Shoin, Tokyo, pp 65–72 (J)

[1] (J): written in Japanese

Hirai M (2014) Korei jinko ido (Migration of elderly population). In: Inoue T, Watanabe M (eds) Shuto ken no koreika (Population aging in the Tokyo metropolitan area). Hara Shobo, Tokyo, pp 53–71 (J)

Hirosima K (2014) Gokei junidoritsu ni yoru sengo todofuken betsu jinko ido no bunseki (An analysis of postwar prefectural migration via total net migration rates). Keizai Kagaku Ronsyu: Shimane Daigaku Hobungakubu Kiyo, Hokeigakka Hen (Journal of Economics: Memoirs of the Faculty of Law and Literature Shimane University) 40:25–44 (J)

Inoue T (1991) Nihon kokunai ni okeru nenrei betsu jinko idoritsu no chiikiteki sai (Regional difference of age-specific migration rates in Japan). Jinbun Chirigaku Kenkyu (Tsukuba Studies in Human Geography) 15:223–250 (J)

Inoue T (2001) Waga kuni ni okeru shogai ido to sono tokusei (Lifetime migration in Japan). Jinko Mondai Kenkyu (Journal of Population Problems) 57(1):41–62 (J)

Inoue T (2002) Jinkogakuteki shiten kara mita waga kuni no jinko ido tenkan (The migration turnarounds in Japan from the demographic viewpoint). In: Arai Y, Kawaguchi T, Inoue T (eds) Nihon no jinko ido: Raifu kosu to chiiki sei (Migration in Japan: Life course and regionality). Kokon Shoin, Tokyo, pp 53–70 (J)

Inoue T (2014) On the mathematical formulation of the cohort cumulative social increase ratio. Working Paper Series, Institute of Economic Research, Aoyama Gakuin University 2014–4:1–18

Ishikawa Y (1994) Jinko ido no keiryo chirigaku (Quantitative geography of migration). Kokon Shoin, Tokyo (J)

Ishikawa Y (ed) (2001) Jinko ido tenkan no kenkyu (Studies in the migration turnarounds). Kyoto University Press, Kyoto (J)

Itoh T (1984) Nenrei kozo no henka to kazoku seido kara mita sengo no jinko ido no suii (Recent trends of internal migration in Japan and "potential lifetime out-migrants"). Jinko Mondai Kenkyu (Journal of Population Problems) 172:24–38 (J)

Kawabe H (1985) Kohoto ni yotte mita sengo Nihon no jinko ido no tokushoku (Some characteristics of internal migration observed from the cohort-by-cohort analysis). Jinko Mondai Kenkyu (Journal of Population Problems) 175:1–15 (J)

Maruyama Y, Oe M (2008) Senzaiteki tashutsusha kasetsu no saikento: Chiikiteki sai to kohotokan sai ni chakumoku shite (Re-examining the "potential out-migrants hypothesis": Focusing on differences among regions and cohorts). Jinkogaku Kenkyu (The Journal of Population Studies) 42:1–19 https://doi.org/10.24454/jps.42.0_1 (J)

Nakagawa S (2005) Tokyo ken o meguru kinnen no jinko ido: Kogakurekisha to josei no sentakuteki syuchu (Selective migration to Tokyo: Female and highly educated). Kokumin Keizai Zasshi (Journal of Economics & Business Administration) 191(5):65–78 (J)

Nakagawa S (2010) 1920–30 nendai no jinko ido to senzaiteki tasyutsusha (Migration in the 1920s–30s and persons potentially expected to leave their birthplaces). In: Takahashi S, Nakagawa S (eds) Chiiki jinko kara mita Nihon no jinko tenkan (Demographic transition of Japan from the viewpoint of Regional Population). Kokon Shoin, Tokyo, pp 193–210 (J)

National Institute of Population and Social Security Research (2013) Nihon no chiikibetsu shorai suikei jinko: Heisei 22 – Heisei 52 (Regional population projections for Japan: 2010–2040). National Institute of Population and Social Security Research (J)

Oe M (1995) Kokunai jinko bunpu hendo no kohoto bunseki: Tokyoken eno jinko shuchu purosesu to shorai tenbo (Cohort analysis of population distribution change in Japan: Processes of population concentration to the Tokyo region and its future). Jinko Mondai Kenkyu (Journal of Population Problems) 51(3):1–19 (J)

Shimizu M (2006) On the quantum and tempo of cumulative net migration. Jinko Mondai Kenkyu (Journal of Population Problems) 62(4):41–60

Shimizu M (2009) Shichosonbetsu no kohoto ruiseki shakai zokahi (Cohort cumulative social increase ratios by municipality: A case study of Nagano Prefecture). Jinkogaku Kenkyu (The Journal of Population Studies) 44:33–42 https://doi.org/10.24454/jps.44.0_33 (J)

Shimizu M (2011) Kokunai jinko ido (Internal migration). In: Ishikawa Y, Inoue T, Tahara Y (eds) Chiiki to jinko kara miru Nihon no sugata (The population geography of contemporary Japan). Kokon Shoin, Tokyo, pp 57–64 (J)

The Japan Institute of Labour (1994) U-tansha ni miru shokugyo to katei seikatsu, chosa kenkyu hokokusho No. 57 (Occupation and family life of return migrants, research report No. 57). The Japan Institute of Labour (J)

Yamaguchi K (ed) (1989) Jinko bunseki nyumon (Introduction to population analysis). Kokon Shoin, Tokyo (J)

Zelinsky W (1971) The hypothesis of the mobility transition. Geographical Review 61:219–249 https://doi.org/10.2307/213996

Chapter 2
Internal Retirement Migration in Japan Revisited

Yoshitaka Ishikawa

Abstract The question of whether retirement migration occurs in Japan has raised a certain concern in the country's population geography since the 1980s. Existing literature has generally taken a rather negative view toward the existence of retirement migration. To reconsider this view, net migration schedules by prefecture/municipality in the country are created using the migration data of the 2010 population census. In particular, the net migration rates of the 60–64 age class after retirement are investigated in detail. Consequently, remarkable human flows to many prefectures/municipalities in peripheral Japan are confirmed; this result indeed indicates positive evidence of retirement migration. If this migration were divided into return migration and non-return migration, it could be inferred that return migration mainly occurs in eastern Japan, while both return migration and non-return migration occur in western Japan. When examined on a municipality basis, it is possible to identify three types of leading destinations for retirement migration: municipalities in Hokkaido, the "Kanto circular villa belt" surrounding Tokyo, and the hilly and mountainous areas in western Japan. Important conditions for attracting retiree migrants include a mild climate, an excellent and scenic natural environment, hot springs, the availability of second houses, nearby airports and highways, and supportive measures for migrants.

Keywords Retirement migration · Leading destinations · Return migration · Second house · Peripheral Japan

2.1 Introduction

Japan's total population peaked in 2008 and then began to decline, a trend that is likely to continue for a long time. Along with the decline in the country's total population, regional differences in population decline have also become an important issue. Much

Y. Ishikawa (✉)
Emeritus Professor, Kyoto University, Kyoto, Japan
e-mail: ishikawa.yoshitaka.86x@st.kyoto-u.ac.jp

attention has been focused on the disparity between peripheral regions, where feelings of impoverishment are growing due to the population having been in decline for some time, and the Tokyo metropolitan area, where the population is continuing to grow and become more concentrated (Masuda 2014). Most municipalities in peripheral regions are suffering from a declining population and net out-migration of young people. In the existing literature, studies on the flows of people heading to peripheral regions have focused on phenomena such as the return migration of people to their hometowns in peripheral regions, *den'en kaiki* (migration to the countryside), the movement of community-reactivating cooperator squad members, and the migration of women concomitant with international marriage (Yamaguchi 2018; Odagiri et al. 2015; Shiikawa et al. 2019; Ishikawa 2011). Another example is retirement migration, which is the change in place of residence triggered by retirement and our focus in this paper.

Previous studies on retirement migration have been conducted overseas (e.g., King et al. 2000; Friedrich and Warnes 2000; Casado-Díaz et al. 2004; Weidinger and Kordel 2015; Gehring 2017) as well as in Japan (e.g., Tahara et al. 2000; Kubo and Ishikawa 2004; Takeshita 2006; Ono 2012). Research on retirement migration has become prominent in Western countries, and it has provided a certain amount of stimulus to promote similar studies in Japan. Retirement migration can be divided into international migration and internal migration based on whether it crosses national borders. The subject of this chapter is internal retirement migration.

The question of whether retirement migration occurs in Japan has raised a certain concern in the country's population geography since the 1980s (Nanjo et al. 1982; Kawabe and Inoue 1991: 166; Otomo 1996: 102–103; Ishikawa 2001: 280–281; Tahara 2007). However, the existing literature has generally taken a rather negative view toward the existence of retirement migration. Such a view is partly due to the limitations of some of the previous studies, specifically their use of questionnaire/interview surveys taken with a small number of respondents, implying the difficulty in determining how well the findings could be generalized. With these limitations in mind, there is a need to perform studies using comprehensive data sources such as the national census, which could achieve thorough coverage.

The purpose of this chapter is to examine whether retirement migration exists by creating net migration schedules of all prefectures/municipalities for the period of 2005–10, scrutinizing their net migration rates of residents in the 60–64 age class and clarifying the actual situation through interview surveys targeting municipalities that have seen a large influx of retirees. By doing so, we intend to reconsider the view of previous studies that have repudiated the existence of retirement migration in Japan. Note that the subject of this paper is Japanese residents. When examining retirement migration by age class in five-year increments, the immediate focus of analysis is on the 60–64 age group, since retirement at age 60 is common in Japan (Ministry of Health, Labour and Welfare 2015). At the time of the 2010 census, this class included the baby boomers born in 1947–49. This generation has a very large cohort size, and the presence or absence of retirement migration among them has been of great interest in the field of population geography.

The structure of this chapter is as follows. Section 2.2 introduces the research method based on migration schedules. Next, in Sects. 2.3 and 2.4, we examine the net migration rates for the 60–64 age group by both prefecture and municipality. Section 2.5 describes the results of interview surveys conducted in municipalities to which significant numbers of retirees have migrated. Finally, the findings of this study are summarized in Sect. 2.6.

2.2 Methodology

In this paper, migration is studied in terms of migration schedules, which are expressed as graphs showing migration rates on the vertical axis and age on the horizontal axis, with a particular focus on net migration in the 60–64 age group immediately after the age of retirement. The advantage of this method is that migration schedules can be generated from comprehensive data provided by the national census. Accordingly, any recorded increase in migration rates among people in the post-retirement age group would provide clear and compelling evidence of retirement migration. Research on migration schedules is one of the key outcomes of the Migration and Settlement research project (1975–82) conducted at the International Institute for Applied Systems Analysis (IIASA) (Rogers and Castro 1986).

Migration rates have been found to vary systematically with age, and they can be expressed structurally as a combination of four components: a pre-labor force component, a labor force component, a post-labor force component, and a constant component. Here, the pre-labor force component corresponds to the age group of people who are living at home with their parents before entering the labor market, and it indicates the rate at which they migrate to other residences together with them. The labor force component indicates the migration rates of people included in the labor market, and the post-labor force component indicates the rate of people after their retirement from the labor market. The constant component is included because even in age groups with low migration rates, there is the possibility that low-level migration could still occur. Of these four components, the post-labor force component reflects the rate of migration triggered by retirement, whereby people exit the labor market.

The IIASA's Migration and Settlement project brought together many eminent researchers in population-related fields from 17 countries and was completed in 1982 (Rogers and Willekens 1986: 5). One of the outcomes of this project was the provision of a post-labor force component in migration schedules, indicating that retirement migration is a common occurrence in many of the countries that participated in this project. Japan also participated in the project, and a report on the domestic migration in the country was published (Nanjo et al. 1982). However, this report does not contain any description of a post-labor force component. In a paper by Rogers and Castro (1986) summarizing research on migration schedules at IIASA, Japan's data were handled by a reduced model that lacks a post-labor force component. The same reduced model was also used by Kawabe and Inoue (1991), who were the first to

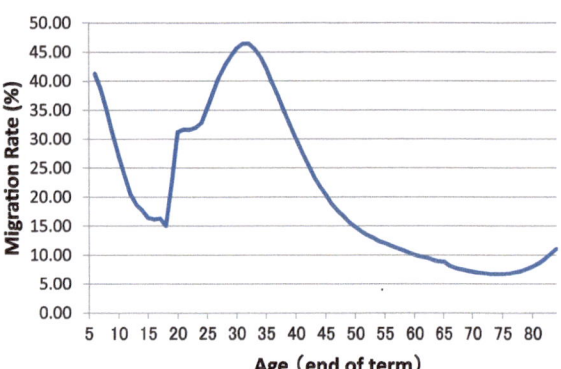

Fig. 2.1 Migration rate in Japan

fully introduce a model migration schedule in Japan and apply it to Japanese data, and by Ishikawa (2001: 207–255), who applied model migration schedules for each Japanese region to a study of the migration turnarounds. The use of such a reduced model seems to assume the viewpoint that retirement migration is not common in Japan.

To begin with, we need to confirm whether retirement migration is observed based on migration data for Japan as a whole. Figure 2.1 shows the national migration schedule based on data from the 2010 Population Census, with the migration rate calculated from the number of people who moved between 2005 and 2010 (excluding those from abroad) in the numerator and the population of the relevant age group in 2010 at the end of this term in the denominator. Here, migration refers to any change in a person's place of usual residence from five years previously according to the national census data, including all categories of migration within the same municipality, migration to a different municipality in the same prefecture, and migration to a different prefecture. In other words, migrants here include all those who changed their address during the above five-year period, regardless of how far they traveled.

Figure 2.1 shows that the migration rate starts to decline from 41.3% at age 5 to a low of 15.0% at age 17. Thereafter, it rises rapidly, reaching a peak of 46.5% at age 31, and then declines almost monotonically to 6.7% at age 73. For people in their mid-70s and older, the rate increases slightly, reflecting their migration to elderly welfare facilities or to live with or near their adult children. At any rate, the figure shows no upsurge in migration rate around the retirement age of 60. This means that retirement migration is still not common in Japan as a whole, as far as can be observed from the 2010 census.

That said, does evidence of retirement migration appear when looking at specific prefectures and municipalities? In this study, the net migration rate is used as the vertical axis of the migration schedule. This is because, although most municipalities in peripheral Japan experience pronounced net out-migration among younger people in their late teens to early 20s, it is possible to visually confirm the extent to which this mobility is recovered by the net in-migration of the retirement age population. Therefore, the following discussion focuses on whether the net migration rate is positive in the 60–64 age group, and whether an upward trend can be seen in the net migration schedule curve.

In calculating the net migration rate for the 60–64-year-old population, it is important to pay particular attention to how the figures for the denominator are prepared. In general, the net migration rate for a particular age group in a particular prefecture or municipality is calculated by dividing the number of net migrations for the period 2005–10 in the numerator by the population number of the relevant age class at the beginning of the term (i.e., 2005) in the denominator. However, in this paper, the denominator is the population in 2010 (i.e., the population at the end of term) minus half the net migration. Ishikawa (2018: 127–128) detailed the specific reasons for this approach, so we do not elaborate upon them here, but basically it provides a value that more accurately reflects the particular migration characteristics of this age group. Note that all of the net migration rates for prefectures and municipalities used below were calculated using this method.

2.3 Examination of Migration Rates by Prefecture

Next, we consider the prefecture-specific results. In the 2010 census report, the migration data for prefectures are listed in age increments of one year, so a net migration schedule can be drawn for each age. The net migration schedules for each age in all 47 prefectures were created, but due to space limitations, Fig. 2.2 only shows the net migration rates for the population aged 60–64.

Japan's geography is often discussed based on the dichotomy between the three major metropolitan areas consisting of 11 prefectures (Saitama/Chiba/Tokyo/Kanagawa, Gifu/Aichi/Mie, and Kyoto/Osaka/Hyogo/Nara) and the peripheral regions consisting of all 36 other prefectures. Only nine prefectures showed a net out-migration rate of the population aged 60–64: Saitama, Tokyo, Kanagawa, Aichi, Kyoto, Osaka, Hyogo, Nara, and Hiroshima. Apart from Hiroshima, these are all included in the metropolitan areas. In particular, the three prefectures of Tokyo, Kanagawa, and Osaka have net out-migration rates greater than 0.75%. By comparison, Aichi (−0.49%), which is the center of the Nagoya metropolitan area, does not have a large net out-migration rate, so it can be concluded that this prefecture is less important as an origin for retirement migration. In contrast, all prefectures in peripheral Japan, apart from Hiroshima, show net in-migration rates. Such spatial variations in the net migration rates indicate the development of widespread flows of retirement migration from the three metropolitan areas to the peripheral regions.

Incidentally, since most retired migrants live on pensions, it can be inferred that they tend to move to areas where the cost of living is lower. Since these conditions are typically found in the rural areas of a country's peripheral parts, retirement migration is generally expected to move from the core of the country to its periphery. Fortunately, the migration effectiveness index is available as a measure of whether migration is bidirectional or unidirectional (Ogasawara 1999: 78–80; Hirai 2007).

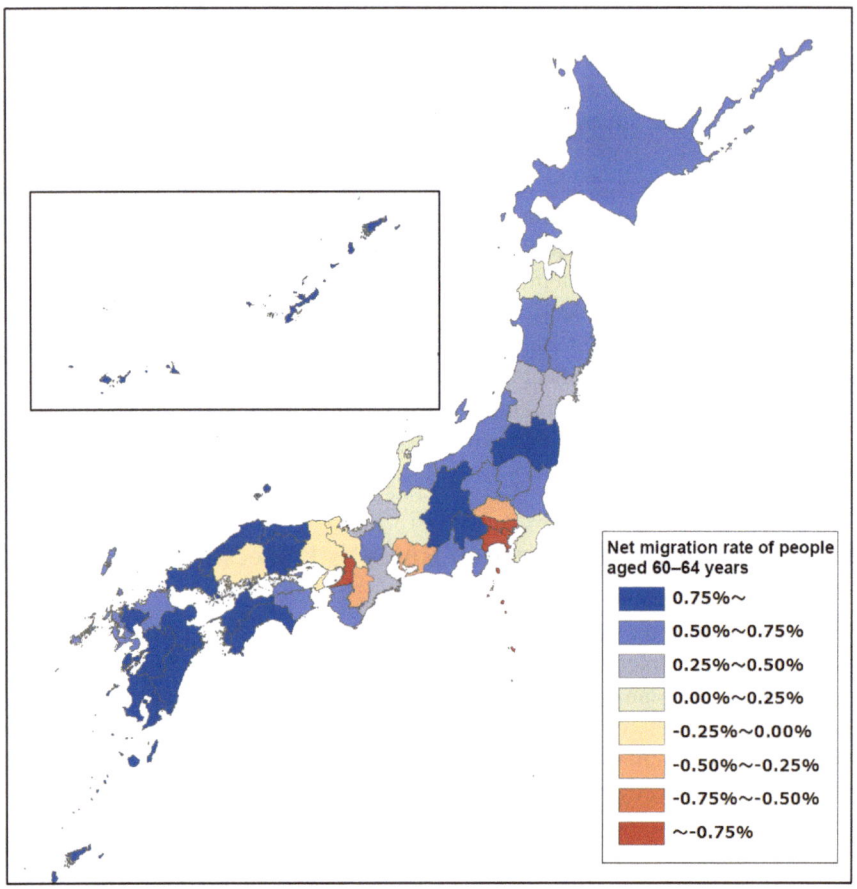

Fig. 2.2 Net migration rate of people aged 60–64 by prefecture

This measure takes a value between 0.0 (inflow and outflow are equal, perfect bidirectionality) and 100.0 (either inflow or outflow is zero, perfect unidirectionality). If this index were calculated using inter-prefectural migration data for each 5-year age group from the 2010 census, its value would be less than 20.0 for the 25–29 through 55–59 age classes, indicating a strong bidirectional trend for these age classes. However, the index value increases from the 55–59 age class and for age 60 and above, and the index value is greater than 27.0 for all age classes, indicating a somewhat stronger unidirectional character. For the 60–64 age class studied in this paper, the value is 37.2. By linking this with the spatial pattern shown in Fig. 2.2, we can confirm again that retirement migration from the three major metropolitan areas to the peripheral regions is dominant.

At the time of the 2010 census, there was net in-migration of the 60–64 age class to rural areas in almost all peripheral prefectures. It is natural to suppose that, since this population includes the baby boomers born in peripheral regions who started migrating to the three major metropolitan areas in the 1960s, some of these migrants wanted to return to their hometowns in peripheral regions after retirement.

A closer look at the positive net migration rate data in Fig. 2.2 reveals that the rate tends to be generally lower in eastern Japan and higher in western Japan. In other words, of the 11 prefectures in the regions of Hokkaido, Tohoku, and Hokuriku, the lowest rate of 0.05% occurred in Aomori and the highest rate of 0.81% occurred in Fukushima, whereas in the 16 prefectures in the regions of Chugoku (except for Hiroshima), Shikoku, and Kyushu/Okinawa, the lowest rate of 0.51% occurred in Fukuoka and the highest rate of 2.25% occurred in Kagoshima, with many cases showing net in-migration rates of at least 1%. How can this pattern of highs in the west and lows in the east be explained? In the western prefectures of Japan, there may be other factors besides a desire for return migration, specifically the warmer climate, and an influx of non-return migration to places other than the individual's hometowns, which has been actively promoted by local governments. To put it another way, it can be inferred that the net migration rate is higher in the prefectures located in Kinki, Chugoku, Shikoku, and Kyushu/Okinawa in western Japan due to a combination of return migration and non-return migration, while the inflows to Hokkaido Prefecture and the prefectures in Tohoku and Hokuriku in eastern Japan are primarily due to return migration. Namely, non-return migration triggered by retirement is mostly toward western Japan, and it seems to be scarce in eastern Japan.

2.4 Examination of Migration Rates by Municipality

Next, we consider the municipality-specific migration rates. As of the 2010 census, there were 1,728 municipalities in the country. The census provides inter-municipal migration data by 5-year age class, from which we were able to draw the net migration schedules for all municipalities. Three municipalities with relatively high net migration rates for the population aged 60–64 are introduced here as examples in Figs. 2.3, 2.4 and 2.5: Shikabe Town in Hokkaido (12.13%), Hokuto City in Yamanashi Prefecture (12.82%), and Kozagawa Town in Wakayama Prefecture (9.56%). Note that since these figures were created with software that automatically sets the range of values on the vertical axis based on the minimum and maximum values, each of these figures has a different vertical axis range.

Shikabe Town is located about 30 km north of Hakodate City, Hokuto City is in the northwestern part of Yamanashi Prefecture, roughly halfway between Kofu City and Lake Suwa, and Kozagawa Town is located at almost the southern tip of the Kii Peninsula. All of these municipalities show a large net out-migration for the 20–24 age class, but in Hokuto City and Kozagawa Town, net in-migration was also observed in the 30s for these municipalities, albeit at a low level. Regarding the 60–64 age class, which is the focus of this paper, these three municipalities show pronounced net in-migration, with a clearly visible increase at the age of retirement.

The net migration schedules for all municipalities were created; however, since it is not possible to show them all here, Fig. 2.6 shows the net migration rates of

Fig. 2.3 Net migration
schedule of Shikabe Town

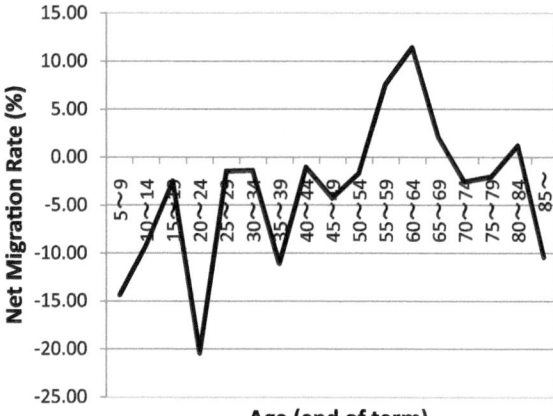

Fig. 2.4 Net migration
schedule of Hokuto City

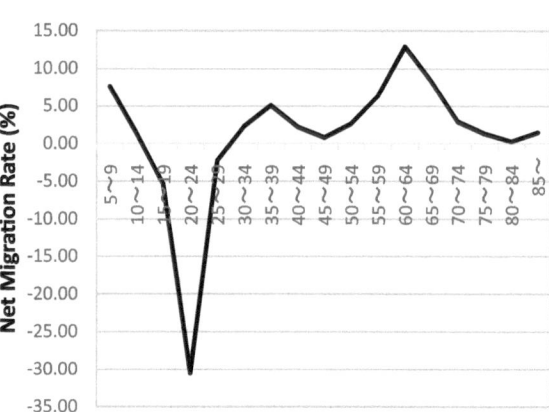

Fig. 2.5 Net migration
schedule of Kozagawa Town

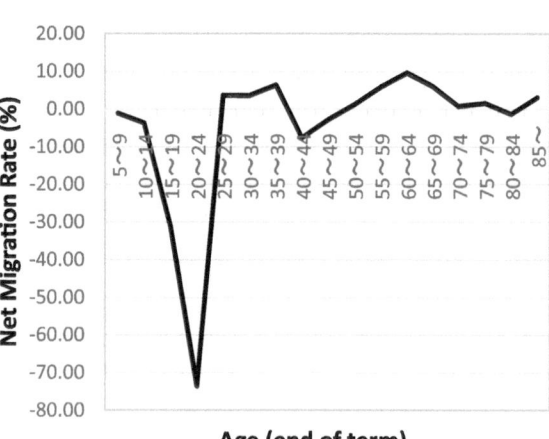

the population aged 60–64 in the form of a map. According to the map, municipalities located in the majority of prefectures in peripheral regions (except Hokkaido) show positive net migration rates. This suggests that there are remarkable flows of retirement migration toward municipalities situated in the peripheral regions. In this map, the positive net migration rate is divided into four classes of 0.0–2.5, 2.5–5.0, 5.0–7.5, and > 7.5%. As a rough guideline, municipalities with a net migration rate of 5% or more can be regarded as significant destinations for internal retirement migration.

As a list of specific municipalities that are important destinations, we also prepared a table of the top 50 municipalities with a population of 3,000 or more at the time

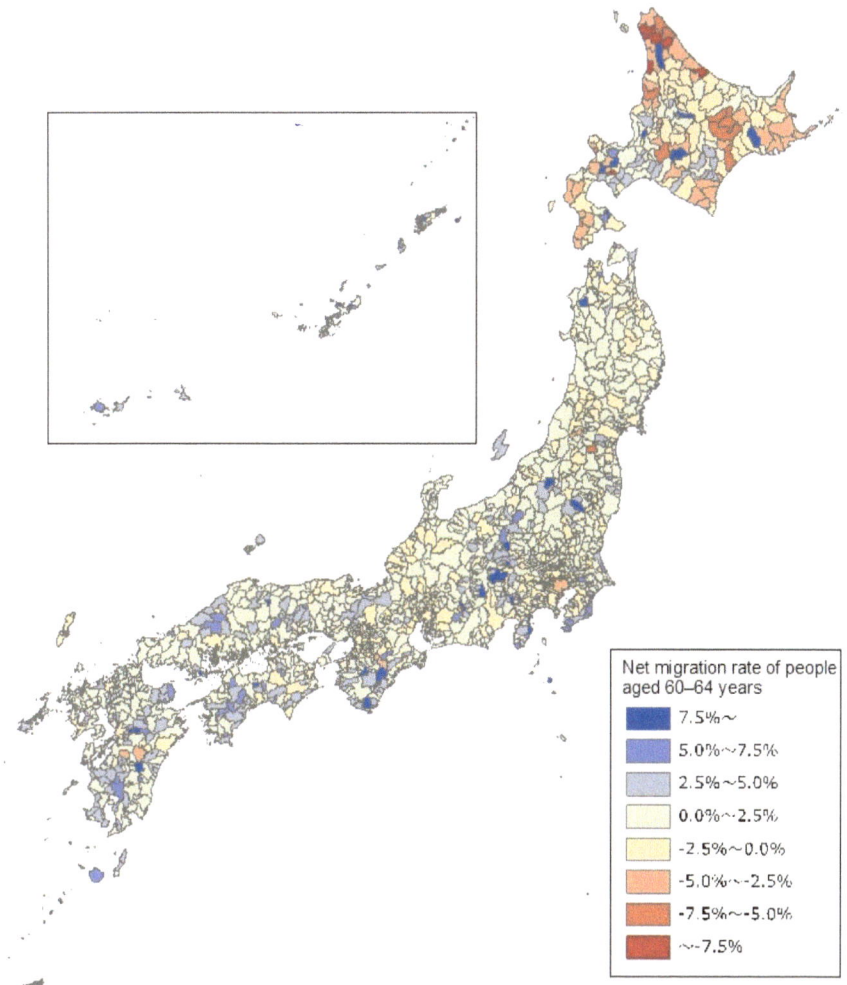

Fig. 2.6 Net migration rate of people aged 60–64

of the 2010 census with the highest net migration rates for the 60–64 age group (Ishikawa 2018: 138), but this is omitted here due to space limitations. Note that Shikabe Town (Fig. 2.3), Hokuto City (Fig. 2.4), and Kozagawa Town (Fig. 2.5) occupy the 8th, 5th, and 14th positions in this table, respectively.

In addition, based on Fig. 2.6, the distribution of retirement migration destinations is examined in further detail. There are hardly any municipalities showing a high net in-migration rate greater than 5% in Tohoku and Hokuriku. Interestingly, such municipalities are distributed in the following three areas that can also be regarded as three types of destinations for significant retirement migration.

First, of the top 50 municipalities in terms of net migration rate for the 60–64 age group, seven are in Hokkaido, namely, Tsukigata Town (15.71%), Kyogoku Town (12.74%), Shikabe Town (12.13%), Niseko Town (10.38%), Higashikawa Town (9.80%), Higashikagura Town (7.59%), and Nanae Town (7.46%). These municipalities are dispersed throughout the prefecture, rather than being concentrated adjacent to one another. Due to the harsh winter climate in Hokkaido, it is thought that many people want to move to Sapporo City after retirement, where it is easier to live. This could be why Hokkaido has so many municipalities with a high negative net-migration rate, as shown in Fig. 2.6. There are several reasons for the high positive net migration rates of the seven municipalities listed above. Shikabe is regarded as one of Hokkaido's best sites for second houses, while Nanae and Higashikagura/Higashikawa are considered suburbs of the cities of Hakodate and Asahikawa, respectively. On the other hand, the popularity of Kyogoku and Niseko could be related to the presence of resorts. Tsukigata showed a significant net in-migration of people not just in the 60–64 age class but across a broad range of ages from the 20–60s, which is due to a prison being located in this municipality.

The second type consists of municipalities surrounding Tokyo, located within a two-hour radius from the city. In clockwise order, these include Yugawara Town (5.86%) in Kanagawa Prefecture, Atami City (8.71%), Ito City (8.94%), and Minamiizu Town (7.33%) in Shizuoka Prefecture, Hokuto City (12.82%) in Yamanashi Prefecture, Hara Village (13.06%) and Karuizawa Town (10.51%) in Nagano Prefecture, Tsumagoi Village (6.84%) in Gunma Prefecture, Yuzawa Town (6.84%) in Niigata Prefecture, Nasu Town (12.69%) in Tochigi Prefecture, Kashima City (6.33%) in Ibaraki Prefecture, and Oamishirasato Town (6.13%), Chosei Village (6.30%), Isumi City (6.54%), and Onjuku Town (18.84%) in Chiba Prefecture. A certain number of these municipalities have villa areas that were developed before the war, and most of them are also well-known as second house areas now (Yasujima and Soshiroda 1991). Shishido (1987: 251) refers to this ring around Tokyo as the "Kanto circular villa belt," and most of the prominent retirement migration destinations in Kanto and Chubu are closely related to the long-standing background of these second-house areas.

The third type is observed in western Japan, west of the Kinki region, where many municipalities with net migration rates above 5% are widely distributed in hilly and mountainous areas and islands. According to Fig. 2.6, such municipalities can be seen in the Kii Mountains, Chugoku Mountains, Shikoku Mountains,

Kunisaki Peninsula, around Mt. Aso, and around Mt. Kirishima. The top 50 municipalities in hilly and mountainous areas include Kozagawa Town (9.56%) and Hidaka Town (6.21%) in Wakayama Prefecture, Onan Town (7.49%) in Shimane Prefecture, Kibichuo Town (6.64%) in Okayama Prefecture, Abu Town (7.22%) in Yamaguchi Prefecture, Higashimiyoshi Town (6.55%) in Tokushima Prefecture, Niyodogawa Town (6.86%) and Yusuhara Town (6.04%) in Kochi Prefecture, Nishihara Village (13.12%), Minamiaso Village (11.72%), Takamori Town (7.59%), and Tsunagi Town (5.96%) in Kumamoto Prefecture, and Hiji Town (5.96%) in Oita Prefecture. On the other hand, municipalities located on islands include Suo-oshima Town (9.25%) in Yamaguchi Prefecture, Yakushima Town (6.44%), Kikai Town (8.06%), and Tatsugo Town (7.19%) in Kagoshima Prefecture, Taketomi Town (6.76%) in Okinawa Prefecture, and municipalities on the main island of Okinawa, including Ogimi Village (10.91%), Nakajin Village (8.98%), Motobu Village (8.32%), Nakagusuku Village (6.81%), and On'na Village (6.66%).

These three types of locations can be identified as important destinations for retirement migration. Of these municipalities that fall into the third category, second-house areas exist, for example, in Kirishima City and Yakushima Town in Kagoshima Prefecture, but such cases are exceptional and rare. Therefore, the third type found in western Japan must be considered as fundamentally different from the second type found in the Kanto and Chubu regions.

The fact that this third category includes leading destinations for retirement migration is probably closely related to the speculation in the above discussion of prefecture-specific migration, in which non-return migration is more common west of the Kinki region. What is important is that the influx of retirees to municipalities as significant retirement destinations in western Japan is thought to play a role in *den'en kaiki* (Odagiri 2014; Odagiri et al. 2015). Such migration is expected to serve as a counterpoint to the perspectives of *chiho shometsu* (disappearance of periphery) (Masuda 2014), which has recently attracted great attention, and the mono-polar concentration in Tokyo.

2.5 Results of Interview Surveys in 12 Municipalities

Regarding municipalities with a pronounced influx of retirees as revealed through the analysis by municipality in the previous section, what specific conditions for this effect can we identify? To clarify the issue, we conducted a series of interviews focusing on municipalities that have high net migration rates for the 60–64 age group and are leading destinations for retirement migration. Specifically, we targeted the 12 municipalities of Date City, Nanae Town, and Shikabe Town in Hokkaido, Nasu Town in Tochigi Prefecture, Isumi City in Chiba Prefecture, Hokuto City in Yamanashi Prefecture, Ito City in Shizuoka Prefecture, Kozagawa Town in Wakayama Prefecture, Suo-oshima Town in Yamaguchi Prefecture, Kunisaki City in Oita Prefecture, Kirishima City in Kagoshima Prefecture, and Ishigaki City in

Okinawa Prefecture. The roughly one-hour interview survey was conducted with the staff of each municipality between August 2014 and May 2015.

The interview included a wide variety of questions, but here we present the results for just four issues: the circumstances and reasons for retirement migration, the degree of satisfaction among retired migrants, the form of permanent or non-permanent migration, and the local municipality's assessment of the influx of retirees.

The first issue relates to how and why these municipalities were chosen as destinations for retirement migration. Responses to this question varied, with some indicating that people had traveled there before and gained a favorable impression and others indicating that people had investigated the area as a potential destination and liked it. There were also many cases of people who owned second houses in the municipality before retiring, and this became a pull factor when choosing where to live after retirement. To summarize the responses from municipalities, pull factors for retirement migration include a mild climate, an excellent and scenic natural environment, hot springs, the availability of second houses, nearby airports and highways, and supportive measures for migrants. The first three conditions can be paraphrased by saying that the amenities must be excellent.

Second, we discuss the satisfaction level of people who have pursued retirement migration. If they are highly satisfied, they will most likely continue to reside in their chosen destination, and if they are highly dissatisfied, they will move away. In other words, if retired movers continue to live in these destinations, it can be safely assumed that their satisfaction levels are high. As a result of this study, it was found that migrants generally tend to remain in their residences for a long time. People only move away in cases where there are compelling reasons, such as the death of a spouse or hospitalization in an advanced health care facility due to failing health. Furthermore, although they may not reach the point of wanting to move away, people can often become dissatisfied if the commercial facilities required for their daily needs are too few or too far away.

Third is the issue of whether this sort of mobility is a permanent migration or non-permanent migration based on the use of two places of residence before and after retirement. Generally speaking, many people who move into municipalities with second houses will naturally enjoy the dual residences because they are economically well off, while many of those who do not have this resource are moving permanently. Permanent movers often live in rented apartments (including public housing) or detached houses (including previously vacant houses). Naturally, permanent migrants are more likely to continue living in the municipality to which they moved. While some people acquire a new cottage at the time of retirement migration, others may inherit it from their parents.

The fourth issue is the local municipality's assessment of the influx of retirees. On the positive side, many cited benefits such as the revitalization due to population growth in the form of inward migration, which helps to mitigate the declining population trend, increases the local allocation of tax grants, boosts the local economy through increased consumer activity, and contributes to the local community by allowing people to make use of their pre-retirement occupations and skills. On the negative side, many also voiced concerns about the increased burden of medical and nursing care costs associated with an elderly population. This is because retirees soon join the ranks of the elderly, and there is concern that their influx could easily cancel out the region's increased subsidies by raising the burden of social security benefits related to medical care and nursing care as they age. As a result, the overall impression is that while local municipalities have a welcoming attitude toward retirement migrants, they are also rather anxious about having to shoulder a heavier burden of social benefits in the future.

2.6 Conclusion

Previous studies have assumed that retirement migration is generally a phenomenon that does not occur in Japan. Based on a cursory examination of national migration rates by age group for the 2005–10 period based on the 2010 census data, it is easy to conclude that retirement migration remains uncommon, as there is no increase in migration rate in the immediate post-retirement years. However, based on an analysis of net migration schedules prepared by prefecture or by municipality and the net migration rate of the population aged 60–64, reflecting retirement age, it can be confirmed that there are remarkable migration flows from the three major metropolitan areas to the peripheral regions. An appreciable number of municipalities have become prime destinations for retirement migration.

An analysis by prefecture reveals that the net migration rate of the population aged 60–64 is higher in the west of Japan and lower in the east. If retirement migration is divided into return migration and non-return migration, it can be inferred that return migration mainly occurs in eastern Japan (Hokkaido and the Tohoku and Hokuriku regions), while return migration and non-return migration both occur in western Japan west of the Kinki region. When examined on a municipality basis, it is possible to identify three types of leading destinations for retirement migration: municipalities in Hokkaido, the "Kanto circular villa belt" surrounding Tokyo, and the hilly and mountainous areas in western Japan. Moreover, important conditions for attracting retiree migrants include a mild climate, an excellent and scenic natural environment, hot springs, the availability of second houses, nearby airports and highways, and supportive measures for migrants. These findings suggest that Japan's internal retirement migration plays a role in "migration to the countryside" and helps to alleviate the mono-polar concentration of population into Tokyo.

Finally, two key points should be addressed. First, this paper is primarily concerned with providing an overview of internal retirement migration but does not confirm the satisfaction levels of the retirement migrants themselves. Nevertheless, interviews in municipalities with high net migration rates in the 60–64 age group suggest that many of the migrants continue to reside in their chosen destinations and that their satisfaction is generally high. Second, although the main source of data for the retirement movements identified in this paper is the 2010 census, which covers the period 2005–2010, similar movements may also have been observed earlier

(Ishikawa 2018; 122, 133). A detailed discussion of when retirement migration first occurred in Japan is a subject for future study.

References[1]

Casado-Díaz MA, Kaiser C, Warnes AM (2004) Northern European retired residents in nine southern European areas: Characteristics, motivations and adjustment. Ageing and Society 24:353–381. https://doi.org/10.1017/S0144686X04001898

Friedrich K, Warnes AM (2000) Understanding contrasts in a later life migration patterns: Germany, Britain and the United States. Erdkunde 54:108–120. https://doi.org/10.3112/erdkunde.2000. 02.02

Gehring A (2017) Pensioners on the move: A 'legal gate' perspective on retirement migration. Population, Space and Place 23:e2007. https://doi.org/10.1002/psp.2007

Hirai M (2007) Koreisha ni yoru todofuken-kan ido no chiikisei (Regional characteristics of inter-prefectural migration by the elderly). In: Ishikawa Y (ed) Jinko gensho to chiiki: Chiri gakuteki apurochi (Population decline and regional imbalance: Geographical perspectives). Kyoto University Press, Kyoto, p. 129–147 (J)

Ishikawa Y (2001) Jinkoido tenkan no kenkyu (Studies in the migration turnarounds). Kyoto University Press, Kyoto (J)

Ishikawa Y (2011) Recent in-migration to peripheral regions of Japan in the context of incipient national population decline. In: Coulmas F, Lützeler R (eds) Imploding populations in Japan and Germany. Leiden and Boston, Brill, p. 420–442.

Ishikawa Y (2018) Ryunyu gaikokujin to nihon: Jinko gensho heno shohosen (New immigration and Japan: Solution to population decline). Kaisei-sha, Otsu (J)

Kawabe H, Inoue T (1991) Jinko ido moderu (Migration schedule model). In: Kawabe H (ed) Hatten tojokoku no jinko ido (Migration in developing countries), Ajia Keizai Kenkyusho, Tokyo, p. 139–170 (J)

King R, Warnes T, Williams A (2000) Sunset lives: British retirement migration to the Mediterranean. Berg, Oxford

Kubo T, Ishikawa Y (2004) "Rakuen"o motomete: Nihonjin no kokusaiintai ido (Searching for"Paradise": Japanese international retirement migration).Jimbun Chiri (Japanese Journal of Human Geography) 56:296–309 https://doi.org/10.4200/jjhg1948.56.296 (J)

Masuda H (ed.) (2014) Chiho shometsu: Tokyo ikkyoku shuchu ga maneku jinko kyugen (Disappearance of peripheral areas: Rapid population decline caused by mono-polar concentration into Tokyo). Chuo Koron Shinsha, Tokyo (J)

Ministry of Health, Labour and Welfare (2015) Heisei 20 nen konenreisha koyo jittai chosa kekka no gaiyo (Overview of the results of the survey on employment of the elderly in 2008). http://www.mhlw.go.jp/toukei/itiran/roudou/koyou/keitai/08/kekka.html. Accessed 30 Nov 2015 (J)

Nanjo Z, Kawashima T, Kuroda T (1982) Migration and settlement: 13 Japan. RR-82-5, International Institute for Applied Systems Analysis, Laxenburg

Odagiri T (2014) Nosanson ha shometsu shinai (Firming and mountain villages will not disappear). Iwanami Shoten, Tokyo (J)

Odagiri T, Fujiyama K, Ishibashi R, Tsuchiya N (2015) Hajimatta den'en kaiki: Genba kara no hokoku (Migration to the countryside that has begun: A report from the field). Nosangyoson Bunka Kyokai, Tokyo (J)

Ogasawara S (1999) Jinko chirigaku nyumon (Introduction to population geography). Taimeido, Tokyo (J)

---

[1] (J): written in Japanese

Ono M (2012) Nihonjin koreisha no kea o motometa kokusai ido: Mareshia ni okeru kokusai taishoku iju to medeikaru tsurizumu no doko kara (Searching for care: International retirement migration and medical tourism in Malaysia among elderly Japanese). Ajia Pashifikku Tokyu (Journal of Asia-Pacific Studies) 18:253–267 (J)

Otomo A (1996) Nihon no jinkoido: Sengo ni okeru jinko to chiiki bunpu hendo to chiikikan ido (Migration in Japan: Population, regional distribution changes and interregional migration in the postwar period). Okurasho Insatsu-kyoku, Tokyo (J)

Rogers A, Castro LJ (1986) Migration. In: Rogers A, Willekens FJ (eds) Migration and settlement: A multiregional comparative study, D. Reidel, Dordrecht, p. 157–208

Rogers A, Willekens FJ (eds.) (1986) Migration and settlement: A multiregional comparative study. D. Reidel, Dordrecht

Shiikawa S, Odagiri T, Sato K, Chiiki Kasseika Senta, Iju Koryu Suishin Kiko (eds.) (2019) Chiiki okoshi kyoryoku-tai: 10 nen no chosen (Community-reactivating cooperator squad: 10 year challenge). Nosangyoson Bunka Kyokai, Tokyo (J)

Shishido M (1987) Karuizawa bessoshi: Hishochi hyakunen no ayumi (History of Karuizawa villa area: A 100-year history of summer resorts). Sumai no Toshokan Shuppan-kyoku, Tokyo (J)

Tahara Y (2007) Intai ido no doko to tenbo: Dankai no sedai ni chumoku shite (Trends and outlook for retirement migration: Focusing on the baby boomer generation). In: Ishikawa Y (ed.) Jinko gensho to chiiki: Chirigaku teki apurochi (Population decline and regional imbalance: Geographical perspectives), Kyoto University Press, Kyoto, p. 43–67 (J)

Tahara Y, Nagata J, Arai Y (2000) Korei kikan ido no katei to sono eikyo ni kansuru kento: Okinawa-ken N buraku no jirei (A study on the process of elderly return migration and its effects: Case of N settlement in Okinawa Prefecture). Ronen Shakai Kagaku (Japanese Journal of Gerontology) 22:436–448 (J)

Takeshita S (2006) Yakushima heno aitan ni okeru chukai fudosan gyosha no yakuwari (The role of a real estate agency in migration to Yakushima). Jimbun Chiri (Japanese Journal of Human Geography) 58:475–488. https://doi.org/10.4200/jjhg.58.5_475 (J)

Weidinger T, Kordel S (2015) German spa towns as retirement destinations: How (pre)retirees negotiate relocation and locals assess in-migration. Two Homelands 42:37–53

Yamaguchi Y (2018) Wakamono no shushoku ido to kyojuchi sentaku: Tokai shiko to jimoto teichaku (Migration trends of young jobseekers and preferred residential location: Heading for a metropolis or staying in a hometown). Kokon Shoin, Tokyo (J)

Yasujima H, Soshiroda A (1991) Nihon bessho-shi: Rizoto no genkei (Japanese villa history: Prototype of the resort). Seiun-sha, Tokyo (J)

Chapter 3
Characteristics of Residential Mobility After the Great East Japan Earthquake: Focusing on Affected Prefectures of the Tohoku Region, Japan

Hirohisa Yamada

Abstract In this study, we overview the residential mobility occurring after the Great East Japan Earthquake in the affected prefectures of the Tohoku region using census data. The characteristics of this residential mobility in these areas are clarified through categorization of the types of mobility identified. Based on the relationships among origin and destination, distance traveled, and the presence or absence of evacuation orders, we categorized the studied mobility into four types: "new built-up area response," "self-exploration," "evacuation," and their "composites." The "new built-up area response type" occurs in response to the construction of new urban areas by the government. The "self-exploration type" is the mobility of individual disaster victims out of the area with the aim of forming a new life. The "evacuation type" is prompted by administrative instructions. In addition, "complex mobility" combines the characteristics of the above three mobility patterns. Residential mobility in the affected areas can be viewed as a process carried out by migrants to recover their daily lives, and the different methods used to secure "safety" for each victim make this mobility complicated and diverse.

Keywords Great East Japan Earthquake · Tohoku region · Residential mobility · Population distribution · Safety

3.1 Introduction

The Great East Japan Earthquake, which occurred at 2:46 p.m. on March 11, 2011, caused extensive damage mainly in the municipalities along the Pacific coast of the Tohoku region. In the affected areas, reconstruction plans have been formulated to ensure the safety and well-being of residents, and the construction of new built-up areas is still in progress.

H. Yamada (✉)
Faculty of Humanities and Social Sciences, Yamagata University, Yamagata, Japan
e-mail: hyamada@human.kj.yamagata-u.ac.jp

© The Author(s), under exclusive license to Springer Nature Singapore Pte Ltd. 2023
Y. Ishikawa (ed.), *Japanese Population Geographies I*,
Population Studies of Japan,
https://doi.org/10.1007/978-981-99-2035-8_3

Along with the government's reconstruction project, there was also a high level of residential mobility of people in the affected area. Clarifying the characteristics of residential migration within the same municipality or across municipalities is an urgent task for reconsidering and implementing urban planning in municipalities with the aim of accommodating disaster victims. Furthermore, this is also essential for developing a manual for recovery planning after large-scale disasters. However, in order to properly characterize the extensive mass residential migration in the big picture, it has been necessary to wait for the collection of data that allows us to quantify it.

This constraint was resolved by the release of the 2015 census results. By comparing the data from the national census conducted in October 2015 (four years and seven months after the Great East Japan Earthquake) with the data from the October 2010 national census (five months before the earthquake), we can examine the population changes and residential mobility that came about as a result of the earthquake.

When considering the areas affected by the Great East Japan Earthquake, it is important to bear in mind that this residential mobility occurred in non-metropolitan areas and, moreover, that the economic vitality of many of the affected municipalities had already been declining for some time due to an ongoing depopulation trend. In municipalities affected by the earthquake, it has become necessary to reorganize built-up areas to suit the needs of residents who chose to remain in the affected areas as well as for residents who took refuge there from other affected areas. This must be done while also ensuring compliance with existing plans from the medium-to-long-term perspective of making these built-up areas more compact. It can be said that the Great East Japan Earthquake presented further challenges to municipalities in the Tohoku region, where the population had already been declining.

The purpose of this study is to use national census data to obtain a bird's-eye view of the residential mobility following the Great East Japan Earthquake that struck the Tohoku region and then to categorize these results in order to clarify the characteristics of residential mobility in the disaster-stricken area.

In this study, we use the term "built-up area" to loosely refer to a district where the population density is relatively high in comparison to the surrounding area, rather than a district with clearly defined absolute values such as a normal densely inhabited district. In addition, we use the term "reorganization" to refer to the spatial transformation process and the future direction of built-up areas due to urban planning aimed at constructing new urban areas after the earthquake. We also use the term "affected prefectures" to refer to the nine prefectures of Aomori, Iwate, Miyagi, Fukushima, Ibaraki, Tochigi, Chiba, Niigata, and Nagano, which have been designated by government ordinance as having suffered particularly severe damage from the Great East Japan Earthquake.

3.2 Demographic Changes Due to the Earthquake

3.2.1 Population Change by Prefecture in the Tohoku Region

Between 2005 and 2010, the population of the six prefectures in the Tohoku region decreased by 3.1%, but between 2010 and 2015 it decreased by 3.8%, indicating that the region's population decline has accelerated. On the other hand, the 2010–2015 period was the first time since the end of World War II that the national population growth rate turned negative (− 0.8%), so the accelerated population decline in the region can be considered a reflection of the national trend.

From the population of people aged 5 years and over in the 2010 and 2015 census data, looking at where people were living five years previously on a per-prefecture basis, we found that in the case of Aomori Prefecture, the proportion of people who were still living in the same prefecture was 94.3% in 2010 and 94.8% in 2015. For the other prefectures in the Tohoku region, the corresponding percentages were 94.6% and 95.2% in Iwate Prefecture, 93.4% and 93.9% in Miyagi Prefecture, 95.0% and 95.3% in Akita Prefecture, 95.1% and 95.6% in Yamagata Prefecture, and 95.0% and 93.3% in Fukushima Prefecture. In Aomori, Iwate, Miyagi, Aomori, and Yamagata, the percentage of respondents who were still living in the same prefecture as five years ago was slightly higher in the 2015 survey than in the 2010 survey. It is possible to identify a similar trend even if the place of residence five years ago is expanded to include the entire Tohoku region. From these results alone, it would be over-simplistic to conclude that these five prefectures experienced greater cohesion within each individual prefecture, or within the Tohoku region as a whole, as a result of the Great East Japan Earthquake. Nevertheless, it is at least clear that the earthquake did not lead to a significant exodus of people from these prefectures or from the region as a whole.

However, in Fukushima Prefecture, the percentage of people remaining in the prefecture itself, or remaining in Tohoku as a whole, compared with five years ago declined after the Great East Japan Earthquake, indicating a progressive outflow of residents to other prefectures and even out of the region. This is a result of the nuclear incident that occurred in the Fukushima Daiichi Nuclear Power Station (abbreviated to F1NPS), which is located in the town of Okuma in Fukushima Prefecture. The analysis results show that this prefecture experienced a unique phenomenon, which cannot be lumped together with the population decline and residential mobility phenomena caused by the earthquake in other prefectures.

3.2.2 Impact of the Earthquake on Demographics

The municipalities affected by the Great East Japan Earthquake are concentrated in the prefectures of Iwate, Miyagi, and Fukushima. Most of the human casualties, including the dead and missing, were caused by the tsunami, which was particularly severe in the coastal areas of the three prefectures.[1] Of the 81 municipalities in the Tohoku region, there were over a hundred human casualties in 26 cities and towns (6 in Iwate Prefecture, 11 in Miyagi Prefecture, and 9 in Fukushima Prefecture).[2]

The number of human casualties exceeded 5% of the 2010 population in Otsuchi Town (8.3%), Rikuzentakata City (7.8%), and Onagawa Town (8.7%). Since human losses are directly reflected in the natural decrease of municipalities, the population change rates of these three municipalities from 2010 to 2015 were − 23.0%, − 15.2%, and − 37.0%, respectively. Between 2005 and 2010, the populations of these municipalities changed by − 7.5%, − 5.7%, and − 6.3%, respectively, so they were already in steep decline before the disaster. Therefore, in these places, which had experienced severe suffering, it seems that the perception of disaster risk and the burden of psychological damage caused by the disaster accelerated the population decline by promoting the exodus of people to other locations.

The relationship between population and the changes in population in municipalities of the six Tohoku prefectures from 2005 to 2010 shows that most of the municipalities experienced a population decrease proportional to their population size (albeit to differing degrees) except Sendai City, Natori City, Tomiya Town, and Rifu Town. However, during the period from 2010 to 2015, in which the earthquake struck, the change in population was unrelated to population size, especially in municipalities that suffered human casualties. In some municipalities, the population increased despite the human suffering, but this is presumably due to disaster victims fleeing heavily damaged municipalities and being accommodated in neighboring municipalities that were less affected.

There were also substantial reductions in the population of the following municipalities in Fukushima Prefecture that occurred as a result of evacuation orders issued in response to the nuclear accident (changes in population from 2010 to 2015 are shown in parentheses): Tamura City (− 4.7%), Minamisoma City (− 18.5%), Kawamata Town (− 7.2%), Hirono Town (− 20.3%), Naraha Town (− 87.3%), Tomioka Town (− 100.0%), Kawauchi Village (− 28.3%), Okuma Town (− 100.0%), Futaba Town (− 100.0%), Namie Town (− 100.0%), Katsurao Village (− 98.8%), and Iidate Village (− 99.3%).

3.3 Characteristics of Residential Mobility in Affected Prefectures

3.3.1 Residential Mobility in Aomori Prefecture

In the 2015 population, those aged five years and over who had lived in their place of usual residence for less than five years at the time of the 2015 census are considered residents who moved after October 2010. In this section, we examine the spatial characteristics of post-earthquake residential mobility by prefecture based on the geographical relationship between the place of usual residence five years ago and that at the time of the census. Of the population aged 5 and over, the proportion of residents who moved during or after October 2010 was 19.7% in Aomori Prefecture, 21.7% in Iwate Prefecture, 25.4% in Miyagi Prefecture, 16.5% in Akita Prefecture, 17.4% in Yamagata Prefecture, and 23.2% in Fukushima Prefecture. Of the affected prefectures, it can be said that Aomori experienced relatively little earthquake-related mobility.

Of the 243,000 migrants who had lived in 10 cities, 22 towns, and 8 villages in Aomori Prefecture five years previously, 130,000 people (53.3%) moved to the same municipality by the time of the census.[3] When we drew movements with a migrant ratio of 5% or more based on the place of usual residence 5 years ago (hereafter referred to as "origin"), we found that mobility with a place of usual residence at the time of the survey (hereafter referred to as "destination"), such as Aomori City, was observed in almost all municipalities in the prefecture, whereas there was a particularly high proportion in the Tosei area. Hachinohe City in the Sampachi/Kamikita region and Hirosaki City in the Seihoku/Chunan region were the predominant destinations of mobility.

The populations of Aomori City, Hachinohe City, and Hirosaki City in 2015 were 288,000, 231,000, and 177,000, respectively, and each functioned as a regional center. Of these three cities, Hachinohe suffered the most earthquake damage in the prefecture in terms of the number of casualties, injuries, and damaged buildings, but like the other two cities, it also became a destination of residential mobility occurring within the region, and we could not find any characteristic that could be attributed to the effects of the earthquake.

3.3.2 Residential Mobility in Iwate Prefecture

Of the 263,000 migrants originating from 14 cities, 15 towns, and 4 villages in Iwate Prefecture since October 2010, 141,000 (53.9%) moved to destinations in the same municipality.[4] However, among the coastal municipalities that suffered extensive damage in the earthquake, over 60% of migrants moved to destinations in the same municipality, such as Rikuzentakata City (67.4%), Yamada City (66.0%), Kamaishi City (65.9%), Ofunato City (63.1%), Tanohata Village (62.8%), and Otsuchi Town (61.1%). In Aomori Prefecture, over 60% of migrants moved to destinations in the same municipality in Hachinohe City (61.7%) and Aomori City (60.4%). The same can be said of Ishinomaki City (67.1%) and Kesennuma City (67.0%) in Miyagi Prefecture and of Iwaki City (63.9%) and Shinchi Town (63.5%) in Fukushima Prefecture. One of the characteristics of Iwate Prefecture is that a high percentage of the population moved within the affected municipalities, including some relatively small municipalities.

Looking at the mobility between municipalities, as pointed out by Morikawa (2016), Morioka City occupies a strongly central position in Iwate Prefecture, and the percentage of migrants from each municipality who ended up in Morioka exceeded 5% with the exceptions of Rikuzentakata City (4.2%), Hiraizumi Town (4.8%), and

Noda Village (4.1%) (Fig. 3.1). On the other hand, over 5% of the migrants from
Morioka City and Ichinoseki City ended up in Sendai City.

Outside of Morioka City, there was mobility between municipalities in the
Kitakami Basin spreading to the south of the city, and in the north of the prefecture,
over 10% of all migrants in Karumai Town and Hirono Town migrated to Hachinohe
City in Aomori Prefecture. In coastal municipalities, mobility along the coastline

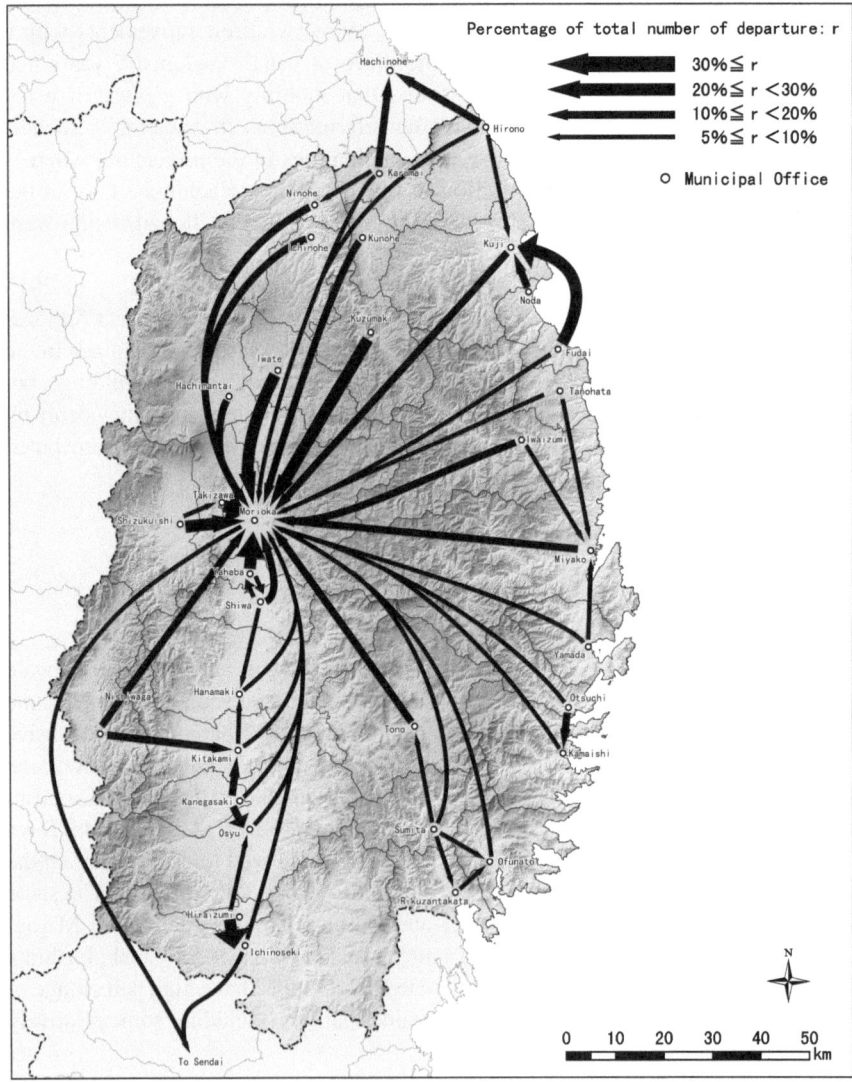

Fig. 3.1 Residential mobility in Iwate Prefecture during 2010–15. Prepared by the author from
National Census data

was extracted in the direction of Kuji City, Miyako City, Kamaishi City, and Ofunato City.

In Iwate Prefecture, the Kitakami Mountains separate the inland area from the coastal area. In addition, the built-up areas of coastal municipalities are highly isolated because they are formed on a narrow plain sandwiched between the Kitakami foothills and the complex ria coastline. These topographical features are particularly noteworthy in the southern coastal area of the prefecture, but in general, it is difficult to move away from the coastal municipalities of Iwate Prefecture while maintaining a daily life.

3.3.3 Residential Mobility in Miyagi Prefecture

Of the 526,000 migrants originating from 13 cities, 21 towns, and one village in Miyagi Prefecture since October 2010, 287,000 (54.5%) moved to destinations in the same municipality.[5] The most significant feature of the residential mobility observed in Miyagi Prefecture is the strong population pull of Sendai City, as indicated by the fact that the ratio of migrants who ended up there exceeded 5% for all municipalities (Fig. 3.2a).

In Sendai City, migration rates exceeding 5% were observed from municipalities in other prefectures. These were Morioka City (5.1%) and Ichinoseki City (5.4%) in Iwate Prefecture, Soma City (5.9%) and Minamisoma City (5.3%) in Fukushima Prefecture, and Ogata Village in Akita Prefecture (6.5%), as well as eight municipalities in Yamagata Prefecture. Sendai City is thus extensively chosen as a destination, and it does not appear to have any topographical barriers to migration. On the other hand, residential mobility to other municipalities was limited to short-distance migration, and we did not extract any north–south movement that bypassed Sendai City (Fig. 3.2b).

In Miyagi Prefecture, as in Iwate Prefecture, the municipalities that were severely damaged by the earthquake are concentrated in coastal areas, but only Kesennuma City and Ishinomaki City experienced movements exceeding 60% of those within their own city. Kesennuma City had a topographically similar environment to the municipalities of Iwate Prefecture. The high mobility rate in Ishinomaki City can be attributed to the fact that this city retained its central role as the prefecture's second-largest city and to the fact that the inland built-up areas incorporated in the municipal mergers of the Heisei era served as a catchment area for the population from the disaster-stricken areas.[6] Both of these cities experienced a sharper population decline in 2010–2015 ($-$ 11.6% in Kesennuma, $-$ 8.5% in Ishinomaki) than in 2005–2010 ($-$ 5.8%, $-$ 3.9%), suggesting that disaster reconstruction projects have not halted the population outflow.

We also detected a high proportion of migratory flows from Onagawa Town and Minamisanriku Town toward Ishinomaki City and Tome City, respectively. This is possibly because the topography of both towns limited the movement of people seeking to evacuate to nearby safe zones.

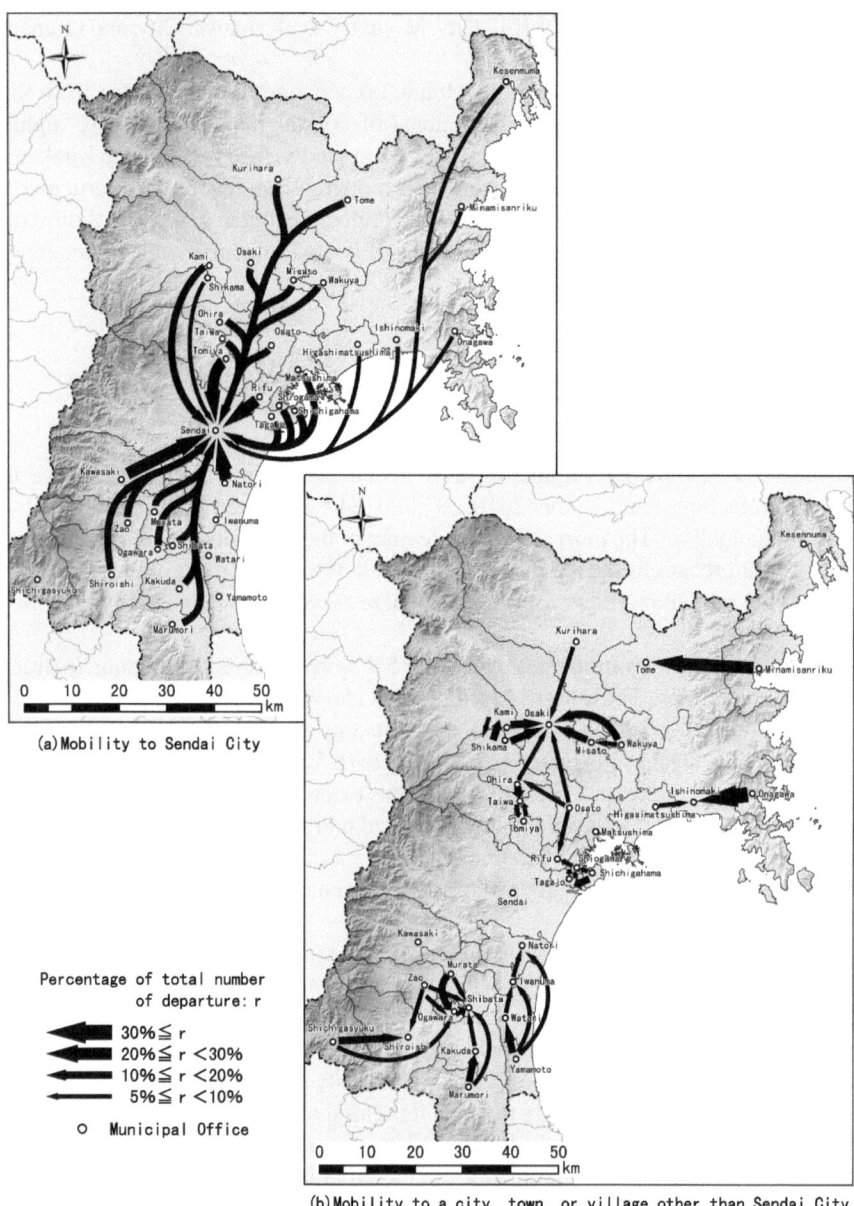

Fig. 3.2 Residential mobilities in Miyagi Prefecture during 2010–15. Prepared by the author from National Census data

3.3.4 Residential Mobility in Fukushima Prefecture

Of the 417,000 migrants originating from 13 cities, 31 towns, and 15 villages in Fukushima Prefecture since October 2010, 183,000 (43.9%) moved to destinations in the same municipality. The corresponding values for the above three prefectures are approximately 53% to 55%, indicating that the ratio of migration within the same municipality was low in Fukushima Prefecture. This is because many of the 59 municipalities in the prefecture are small in area, which meant that more people had no choice but to move to a municipality out of the region as a result of evacuation orders.

Cases where there was a mobility rate of 5% or more from the origin corresponded to migration from 11 municipalities to Iwaki City, from 23 municipalities to Koriyama City, from 7 municipalities to Fukushima City, and from 17 municipalities to Aizuwakamatsu City (Fig. 3.3). Of these, Iwaki City had a population of 350,000 in 2015, and it is the prefecture's largest city, just under Sendai City within the Tohoku region as a whole. Koriyama City (335,000) is located nearly at the center of Fukushima Prefecture and is its second largest city by population, Fukushima City (294,000) is the prefectural capital, and Aizuwakamatsu City (124,000) functions as the center of the Aizu region.

As is clear from these movements, population size is not a straightforward indicator of population attraction within Fukushima prefecture. The Echigo Mountains, the Ou Mountains, and the Abukuma Highlands, which make up the topography of the prefecture, have greatly influenced the historical development of each municipality and have defined the formation of its cultural spheres and transportation networks. It seems that the isolated nature of the seven regions into which Fukushima Prefecture is divided has resulted in each of these regions developing a cohesive character, and thus residential mobility tended to take place within these regions.[7]

However, migrations from Hirono Town, Naraha Town, and Futaba Town to Iwaki City, Tomioka Town, Kawauchi Village, Koriyama City, and Okuma Town to Aizuwakamatsu City, Namie Town to Nihonmatsu City, and Katsurao Village, Miharu Town, and Iidate Village to Fukushima City were due to the collective relocation from towns and villages that were ordered to evacuate in response to the nuclear power plant accident and also due to the need to relocate their town hall functions. The high rate of migration to Iwaki City among towns and villages in coastal areas is thought to be due to the additional migration arising from the continuity of daily living areas.

Although residents gradually returned to their homes when the evacuation order was lifted and business resumed at the town halls, the 2015 population has not yet significantly shown this trend.

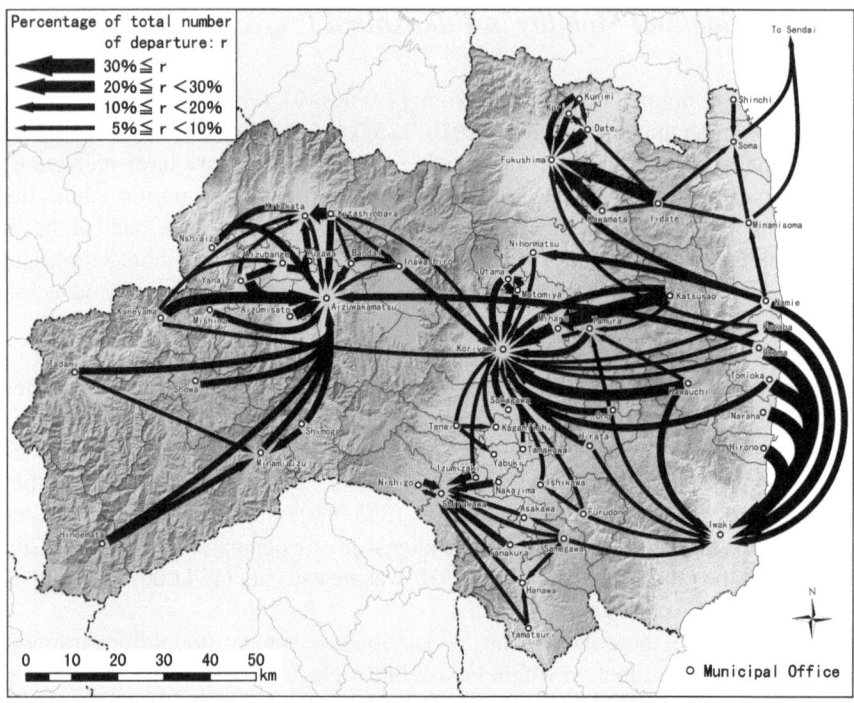

Fig. 3.3 Residential mobilities in Fukushima Prefecture during 2010–15. Prepared by the author from National Census data

3.4 Changes in Population Distribution

3.4.1 Population Distribution of Coastal Municipalities in Iwate Prefecture

The analysis so far has revealed that, among the prefectures affected by the disaster (recall that Aomori Prefecture showed relatively little mobility), the residential mobility observed in Iwate, Miyagi, and Fukushima shows features that indicate disaster as its caused. In this section, we use 500-m mesh data from the national census to identify changes in the population distribution of these three prefectures. However, if the population distribution of an entire prefecture were mapped on a 500-m mesh, it would be difficult to show any feature in terms of the scale. Therefore, in the following discussion, we extract areas where major characteristics can be identified at the same scale.

As a case of coastal municipalities in Iwate Prefecture, we consider Kamaishi City, Otsuchi Town, and Yamada Town. The population change rates of these three municipalities during 2010–15 were − 7.0%, − 23.0%, and − 15.0%, respectively, indicating significant declines in population. It should be noted that the changes in population distribution profiles shown below do not reflect a process of absolute growth due to industrial development and population growth but rather a process of built-up areas being reorganized by the remaining residents.

The population distributions in these municipalities show that the overall population density was already low in 2010, before the earthquake (Fig. 3.4). Rather than the planned formation of built-up areas with concentric structures centered on major urban facilities, the city and the towns arose from long and narrow village-sized built-up areas that formed spontaneously along the rivers flowing into the small inlets of the ria coastline. Each built-up area is topographically hemmed in, and there is little continuity between them.

The main industrial base of coastal municipalities in the prefecture was fishing, and areas with a relatively high population density were also formed in the coastal areas, but most of these were destroyed by the tsunami, causing great human suffering. Therefore, from the viewpoint of disaster prevention and mitigation, few municipalities have attempted to reorganize their built-up areas in the same locations. By comparing the 2010 and 2015 population distributions in Yamada Town, Otsuchi Town, and Kamaishi City, we can see that all of their built-up areas have been moving inland. In the previous section, we showed that the number of migrants from Otsuchi Town to Kamaishi City constituted between 10 and 20% of the total out-migrants

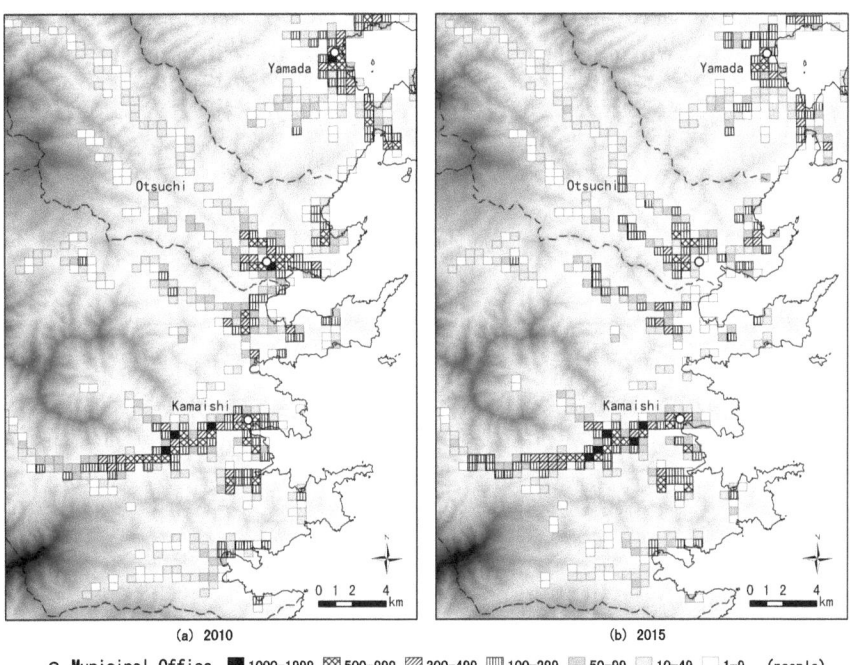

(a) 2010   (b) 2015

○ Municipal Office ■ 1000-1999 ▨ 500-999 ▨ 300-499 ⦀ 100-299 ▨ 50-99 ▨ 10-49 ☐ 1-9   (people)

Fig. 3.4 Change in population distribution of coastal municipalities in Iwate Prefecture. Prepared by the author from National Census 500-m mesh data

from the town (Fig. 3.4). However, when we compare the population distributions of 2010 and 2015, it cannot be said that spatial continuity of the long and narrow built-up areas formed along the small rivers increased after the earthquake.

3.4.2 Population Distribution of Coastal Municipalities in Miyagi Prefecture

Looking at Sendai City and its adjacent northeastern municipalities in Miyagi Prefecture, the built-up areas are continuous, in contrast to the coastal areas of Iwate Prefecture. The mesh-based population change from 2010 to 2015 shows a significant population decrease in coastal areas, showing the impact of the tsunami damage (Fig. 3.5). Although this consequently transformed the population distribution, we did not observe the shift of a densely populated district as seen in the case of Iwate Prefecture. In built-up areas beyond a certain population size, the city center tends to be fixed, making it difficult to rebuild the city center in another location even if it has suffered catastrophic damage. In addition, from the viewpoint of compact city policies and environmental protections set forth in pre-earthquake city planning policies, requests for new residential development after the earthquake were kept to a minimum (Yamada 2018). This may also have played a role in maintaining the basic structure of built-up areas.

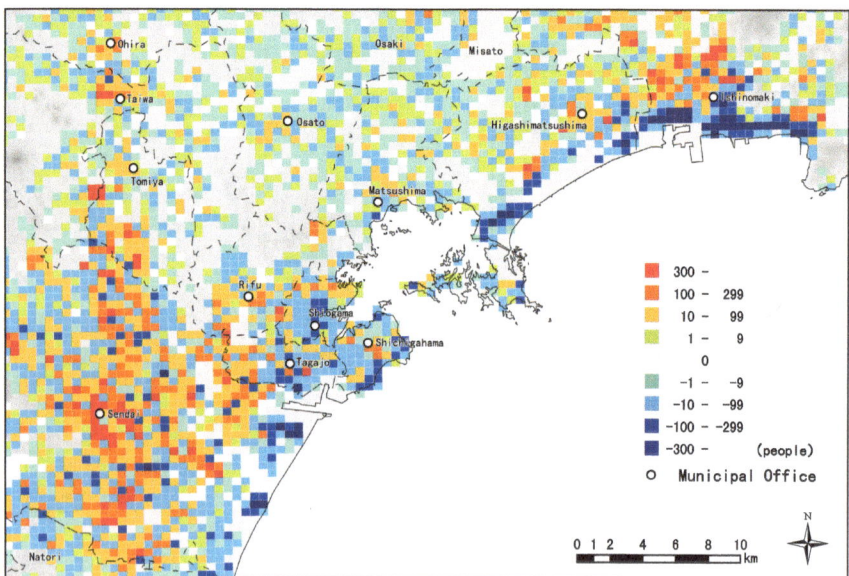

Fig. 3.5 Population change in coastal municipalities in Miyagi Prefecture during 2010–15. Prepared by the author from National Census 500-m mesh data

In Sendai City, the gaps in densely populated areas were filled between 2010 and 2015, and a remarkable population increase also occurred in the suburbs, filling and expanding built-up areas. Population growth in Sendai City also spread to the surrounding municipalities, including Tomiya Town, Taiwa Town, and Ohira Village, which had been developed as dormitory towns to the north of the city, and Rifu Town and Tagajo City to the northeast of the city.

In 2015, the population of Miyagi Prefecture was 2.33 million, and that of Sendai City was 1.08 million. Sendai City functions as the primary city not only of Miyagi Prefecture but also of the entire Tohoku region, and its population had been increasing even before the earthquake. Housing continued to be built in the city to accommodate its growing population, and it is presumed that there was a large stock of affordable (low-rent) old houses. Therefore, it would have been possible for disaster victims to move faster and at lower cost compared with building a new house. It can be inferred that the influx of disaster victims from the affected areas in the prefecture, combined with the existing population concentration in the city, promoted the distribution of housing stock, leading to population growth in existing residential areas across the city center and suburbs.

3.4.3   Population Distribution of Coastal Municipalities in Fukushima Prefecture

In the pre-earthquake population distribution of Iwaki City and the municipalities to its north in Fukushima Prefecture, there is little east–west continuity due to the presence of the Abukuma Highlands, but there is substantial north–south continuity along the straight coastline (Fig. 3.6a).

However, after the earthquake, the entire municipal areas of Tomioka Town, Okuma Town, Futaba Town, Namie Town, Katsurao Village, and Iitate Village were designated as Alert Areas or Planned Refuge Areas due to the nuclear accident. As a result, almost all of the residents left their own towns and villages, and the population density of land that had previously formed built-up areas has dropped to zero (Fig. 3.6b).[8]

Hirono Town is located within 20–30 km of the Fukushima Daiichi Nuclear Power Plant and was initially designated as an Emergency Evacuation Preparation Zone. Naraha Town is located within 20 km of the plant and was largely designated as an Alert Area. Tamura City and Kawauchi Village were partially designated as Alert Areas. In these municipalities, the evacuation orders were either lifted or eased in September 2011, September 2015, April 2014, and October 2014, respectively. As a result, the town governments of Hirono, Naraha, and Kawauchi, which had set up temporary offices in other cities, were able to resume operations at their respective town halls, and residents began to return. More evacuation orders were lifted after 2015, and as of May 2019, Futaba Town was the only municipality where the entire

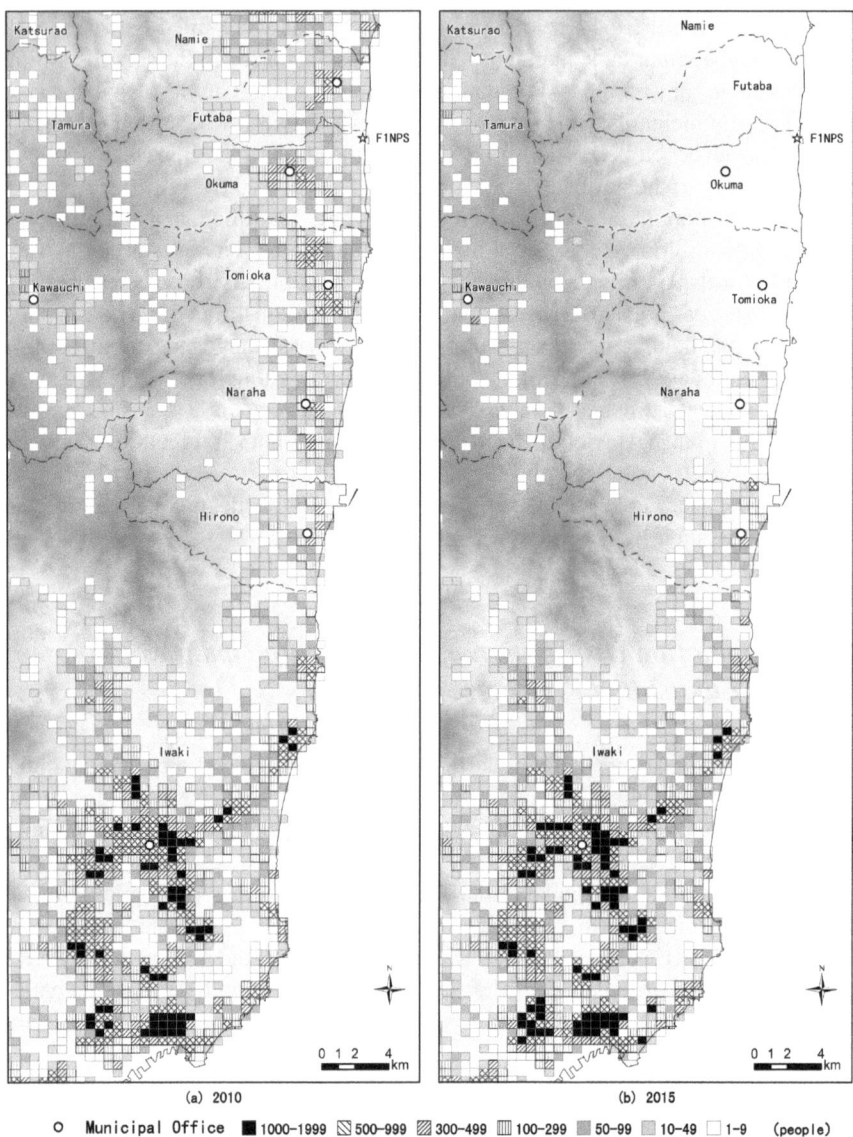

Fig. 3.6 Change in population distribution of coastal municipalities in Fukushima Prefecture. Prepared by the author from National Census 500-m mesh data

area was still under evacuation orders. However, seeing the 2015 population distribution of Hirono Town, where the evacuation order was lifted after approximately six months, it cannot be said that the situation has restored to what it was before the earthquake. In addition, in Okuma Town, where the existing urban area is likely to remain subject to evacuation orders for some time yet, the town hall has been relocated for resumption of business to an area where the evacuation order has been lifted, and it is expected that new urban areas will continue to be constructed while most of the pre-disaster built-up areas remain non-functional.

Iwaki City, which functions as the regional center of the coastal area of Fukushima Prefecture, has become the location of multiple temporary offices and branch offices of affected municipalities, and it appears to have been a preferred destination for individuals searching for places suitable for relocation. Its population grew by 2.3% in 2010–2015. However, in this city, which recorded a population decline of − 3.5% between 2005 and 2010, the housing market was not as active as in Sendai City, and the supply of housing in and around existing residential districts is interpreted as a rapid response to the sharp increase in demand for housing following the earthquake.

3.5 Categorization of Residential Mobility in Affected Areas

3.5.1 New Built-Up Area Response Type

For the residential mobility of disaster victims, re-housing within the same municipality is the first option to consider, followed by moving to neighboring municipalities. The stress felt by the victims can be greatly reduced by maintaining their daily living area, but at the same time, if they want to eliminate the anxiety of re-exposure to signs of local disaster, they might choose living further inland or in houses that are objectively considered to have a lower disaster risk. This results in mobility that responds to urban reorganization plans formulated and implemented by the municipality in question. In this study, we refer to this as "new built-up area response type" mobility.

"New built-up area response type" mobility tends to be influenced by community factors such as deep territorial or family ties in the former place of residence, which make it difficult to move to another area, and topographical factors that cause people to become highly segregated. In addition, disaster victims who prioritize the maintenance of their daily living areas have a high level of community awareness. If they can reside in new built-up areas, they will feel more secure in moving there and will also take less time to complete their migration. They will provide the means to support the construction of new built-up areas, making their intentions better reflected in the reorganization of these areas.

Of the cases discussed in this study, the mobility observed within and between municipalities in the coastal areas of Iwate Prefecture and northeastern Miyagi

Prefecture corresponds to this type of mobility. The changes in the shape of built-up areas such as those observed in Kamaishi City and Otsuchi Town can be considered the result of a match between the wishes of regional residents and the plans for the reorganization of built-up areas toward locations that are further inland, taking disaster prevention and mitigation into consideration.

3.5.2 Self-exploration Type

Following a disaster, if the people it has affected seek to achieve "peace of mind" in terms of work or education, as well as "convenience" in their daily lives, they should target their search for new residences to municipalities that are higher economically than the municipality in which they currently live. Since most of the affected municipalities have long been noted for their economic inactivity, it seems that the target municipalities of search behavior are not the neighboring municipalities of comparable size but municipalities of at least a sufficient rank to function as regional centers.

This sort of mobility was not dictated by someone else but resulted from decisions made by individuals based on information obtained through their personal search activities. Although these are basically individual migrations, there are only a limited number of cities with an abundance of opportunities for resuming work and educational activities and where a convenient lifestyle can be attained after a short period of search activity. As a result, these migratory flows are considered to have occurred to the extent that they could be extracted as a phenomenon. The migration to Sendai City, which has overwhelming population attractions, falls into this category, which we call "self-exploration type" mobility in this study. Note that when migrants chose not to remain within their daily life sphere, travel distance was no longer a meaningful constraint. This resulted in short- to long-distance mobility to some municipality as a fixed destination.

Since disaster victims must obtain affordable housing very quickly, they are forced to conduct search activities that are disadvantageous in terms of time and budget compared with other searchers. As a result, they tend to move into second-hand properties and old ready-built houses that have been on the market for a long time. This is the reason why gaps in the densely populated areas of Sendai City have been filled and the built-up areas have expanded into the suburbs. In other words, the self-exploration behavior of disaster migrants stimulated several segments of the housing stock market, which had been stagnant before the earthquake, leading to revitalization of the housing market as a whole.

From the viewpoint of spatial structure, "self-exploration type" mobility tends to result in intensive migration from multiple affected municipalities to a single municipality, which would affect the restructuring plans of the built-up areas in the destination municipality. However, these migrants are preoccupied with recovering their own lives, and thus any collectivity is weak. Therefore, this mobility did not lead to structural changes in the built-up area, such as the formation of new residential estates, but only to filling in the gaps in densely populated districts and increasing the densely populated districts in the same area.

3.5.3 Evacuation Type

Residential mobility caused by the nuclear accident took place under a tense situation, where short-term safety was prioritized over medium/long-term well-being. It can be said that the affected people did not move of their own volition but were forced to move by administrative evacuation orders. Since this evacuation was necessary to avoid damage to health from residual radiation, this resulted in long-distance mobility that greatly exceeded normal travel distance within the daily living areas of residents up to that time. In this study, we refer to this as "evacuation type" mobility.

In municipalities where evacuation orders were issued, town hall functions were also moved to other municipalities. Residents who were forced to move to distant areas took refuge in temporary housing prepared near temporary town halls. It was clear that evacuation life would be prolonged, and many residents wanted to move from the temporary housing, but it was difficult to find housing in an unfamiliar area. In addition, their search activities were likely to have been severely limited by the need to maintain proximity to the town hall in order to process their evacuation; for evacuees, the town hall is the most important facility in their evacuation. Therefore, it was expected that the return of residents would proceed along with the return of the town hall functions.

However, according to a survey of residents' intentions conducted in October 2018 by the town of Namie, whose town hall functions had been restored in March 2017, only 11.8% said they wanted to return, 30.2% said they had not yet made a decision, and 49.9% said they had decided not to return.[9] With nearly half of the residents having already lost the will to return to their homes, and about a third unable to make a clear decision even seven years after the disaster, it is necessary to conduct a fundamental review of the functions and structure of former built-up areas whose population decreased to zero as a result of mass evacuation. The more time passes, the greater the physical devastation of these former residential areas, making their former residents even less willing to return. Kitamura and Moritomo (2016) analyzed the return of residents to Kawauchi Village from materials provided by the village itself. They noted that the village's rate of elderly population jumped from 34.0% to 40.3% between March 2011 and June 2015, indicating that the ages of returnees were biased toward the elderly. It is said that younger people are more likely to suffer from the effects of residual radiation than the elderly, so households with younger children are probably less likely to have a desire to return.

3.5.4   Composite Type

Migrations with the destination of Iwaki City, Fukushima Prefecture, are considered to be a mixture of "new built-up area response type" mobility, originating within the city itself, and "self-exploration type" and "evacuation type" mobilities that brought people into the city from elsewhere. However, each of the arrows mapping to Iwaki City in Fig. 3.3 cannot be strictly classified into any of these three mobility patterns. As a phenomenon, the mobility flows that arrive in this city combine the characteristics of each of the other three mobility patterns. Here, we refer to this as "composite type" mobility. A similar form of mobility can also be observed for the migrations of people that ended up in Fukushima City or Koriyama City.

In the case of Iwaki City, the earthquake was followed by a rapid increase in "composite type" mobility for various reasons, which is thought to have been the cause of a major positive transformation in the city's demographics. However, "composite type" mobility is a mixture of migration patterns, each with its own characteristics, and the inability to identify a general trend is a problem for a city's urban reorganization. Even if the population increase is temporary, the city's urban reorganization plans will probably require detailed revisions while keeping an eye on the trends of residents. Although the same points could be made with regard to all of the affected municipalities, these are issues that require particular attention in municipalities where "composite type" mobility is prominent.

3.6   Conclusion

In this study, we clarified the characteristics of residential mobility in the affected prefectures of the Tohoku region through a bird's-eye view of the residential mobility after the Great East Japan Earthquake and the related mobility patterns.

In general, an individual residential migration is completed when the migrant has resolved the issues experienced prior to moving, as a result of the need to move, and has attained a certain level of satisfaction. On the other hand, residential mobility that occurs in a disaster-affected area is caused by external factors that are difficult to predict, such as earthquakes. Decisions that have to be made in the chaos following a disaster are tentative, and the affected migrants would then need to take a certain amount of time to calmly reconsider their decisions. Therefore, it is rare for a migrant's wishes to be fulfilled on the first move, and there are cases where multiple moves are repeated. This is a major difference from a ordinary residential move.

In our categorization of residential mobility in the affected areas, "new built-up area response type" mobility is a short-term, short-distance movement of the affected residents within their own municipality, where their mobility is incorporated into the urban area reorganization plan of the local government. However, "self-exploration type" mobility is not responsive to the urban reorganization plans of the municipalities from which people move because they often move out of the area. Even in the municipalities to which people move, they feel the need to prioritize establishing their own livelihood infrastructure before developing any sense of belonging to their destination municipality. On the other hand, there are fears that "evacuation type" mobility caused by the nuclear accident will reduce the willingness of migrants to return due to concerns about health risks and the physical devastation of built-up areas, leading to a significant population decline in their original municipalities.

The important point is when and where the sense of belonging solidifies for disaster victims in any mobility pattern. Those who have lost their homes have no choice but to secure peace of mind by moving elsewhere. Therefore, they may continue to engage in secondary and tertiary mobility until they achieve peace of mind. Although reconstruction projects are still ongoing in affected areas due to the magnitude of this unprecedented earthquake, it is necessary to address the issue of victims who have not yet been able to achieve peace of mind and to complete the process of residential mobility caused by the disaster as soon as possible.

Notes

(1) Our understanding of the damage caused by the earthquake in this study is based on the report "About the Tohoku Region Pacific Coast Earthquake (Great East Japan Earthquake) (158th report)" by the Fire and Disaster Management Agency. The report includes deaths related to the earthquake disaster.

(2) Fujisawa Town in Higashiiwai District existed at the time of the earthquake but was merged with Ichinoseki City in September 2011. Therefore, the damage in Fujisawa Town is included with that of Ichinoseki City.

(3) Percentages are calculated using real numbers, not the round numbers shown in the text. The same applies to Iwate, Miyagi, and Fukushima prefectures.

(4) As remarked in Note 3, Fujisawa Town was merged with Ichinoseki City, so the number of municipalities as of 2015 is shown here.

(5) Tomiya City became a city in October 2016, but in 2015 it was still Tomiya Town, so in the number of municipalities, it is listed as a town.

(6) On April 1, 2005, the towns of Mono, Kanan, Kahoku, Kitakami, Ogatsu, and Oshika were merged into Ishinomaki City to form a new and larger Ishinomaki City.

(7) According to the Fukushima Prefecture website (https://www.pref.fukushima.lg.jp/sec/011 45a/yakuba.html#kenchu, accessed July 15, 2019), the prefecture consists of the Soso region and Iwaki coastal regions east of the Abukuma Highlands, the Kempoku, Kenchu and Kennan regions, which are sandwiched between the Ou Mountains and Abukuma Highlands, and the Aizu and Minamiaizu regions, which are surrounded by the Echigo Mountains and Ou Mountains.

(8) As of October 2015, six towns and villages were still subject to evacuation orders, and the 2015 populations of Tomioka Town, Okuma Town, Futaba Town, and Namie Town were zero, while the populations of Katsurao Village and Iitate Village were 18 and 41, respectively.

(9) Namie Town website (https://www.town.namie.fukushima.jp/soshiki/2/namie-factsheet.html, accessed July 15, 2019).

References[1]

Fire and Disaster Management Agency (2019) Heisei 23 nen (2011 nen) Tohoku chiho taiheiyo-oki jishin (higashi nihon daishinsai) ni tsuite, dai 158 ho (2011 off the Pacific coast of Tohoku Earthquake (No. 158)). https://www.fdma.go.jp/disaster/higashinihon/items/158.pdf. Accessed 17 Aug 2022 (J)

Kitamura I, Morimoto Y (2016) Kison-ritsu to gyosei-ku tono kankeisei ni kansuru chosa hokoku: Fukushima-ken Futaba-gun Kawauchi-mura no jirei kara (Research on the relation between rate of permanent return and community activities: Case of Kawauchi Village, Fukushima Prefecture). Fukushima Daigaku Chiiki Sozo (Journal of Center for Regional Affairs, Fukushima University), 27(2):52–60 (J)

Morikawa H (2016) 2010 nen no jinkoido karamita nihon no toshi sisutemu to chiki seisaku (An overview of Japanese urban systems based on an analysis of internal migration in 2010 and regional development policies). Jimbun Chiri (Japanese Journal of Human Geography), 68(1):22–43. https://doi.org/10.4200/jjhg.68.1_22 (J)

Yamada H (2018) Hisaichi ni okeru tochi riyo kaihen no jittai to kongo no kadai: Miyagi-ken ishinonami-shi wo jirei to shite (The actual situation and future issues of land use change in the disaster area: A case study of Ishinomaki City, Miyagi Prefecture). In: The Tohoku Geographical Association (ed) Higashi nihon daishinsai to chirigaku (Great East Japan Earthquake and geography). Sasaki shuppan, Sendai, p 91–102 (J)

---

[1] (J): written in Japanese

Chapter 4
Spatial-Cycle Model Phases and Differential Urbanization of Cities in the Era of National Population Decline: Japanese Cities 1980–2015

Hyogo Kanda, Yuzuru Isoda, and Tomoki Nakaya

Abstract This chapter reports on the transitions in the urban structure of 109 cities in Japan based on the constituent municipal population of their functional urban regions from 1980 to 2015. The cities consist of central and suburban municipalities and are categorized based on changes in the population of each of these two types of municipalities by applying the spatial-cycle model (SCM). In the 1980s and 90s, many cities in Japan experienced urbanization and suburbanization. In these decades, the largest cities had been showing a tendency of decentralization. However, after 2000, most cities showed a tendency toward centralization, more so among the smaller cities. This result was different from the hypothesized transitions of SCM, lacking the disurbanization phase and often progressing in the reverse direction. In Japan, a clear positive correlation between existing population size and growth has emerged as economic and social changes have proceeded, such as the changing industrial structure and the declining birthrate along with accelerated aging. Population decline occurred earlier in smaller cities than in the larger cities, as did the changes in urban structure, contrary to the predictions of differential urbanization theory. Since the populations of the largest cities are also likely to decrease soon, it is very likely that that these larger cities will also become more centralized as they follow the same tendencies of structural changes that smaller cities have recently experienced.

Keywords Spatial-cycle model (SCM) · Differential urbanization · Compact city · Population decline

H. Kanda
Ministry of Land, Infrastructure, Transport and Tourism, Tokyo, Japan

Y. Isoda
Graduate School of Science, Tohoku University, Sendai, Miyagi, Japan

T. Nakaya (✉)
Graduate School of Environmental Studies, Tohoku University, Sendai, Miyagi, Japan
e-mail: tomoki.nakaya.c8@tohoku.ac.jp

4.1 Introduction

Japan has experienced a decline in its total population since 2008 (IPSS 2018), and even its cities, traditionally viewed as spaces where people live or work in high concentrations, are beginning to lose population. The same applies to large conurbations, and the population of the wards of Tokyo is expected to reach a peak around 2030 and to decline thereafter (TMG 2016). Under these circumstances, it can be difficult to maintain various infrastructures and public services if the urban structure remains diffuse, so the national government and some local governments are promoting Location Optimization Plans based on the idea of compact cities, where urban functions are concentrated in city centers (MLIT 2018).

But in many cases, local governments have tacitly allowed suburban development to take place despite having formulated Location Optimization Plans, and this issue has become a social concern (Nikkei 2018). In addition, Kamezawa (2010) pointed out that the problem of declining city centers and shopping districts still exists in the 2000s, and according to Ujihara et al. (2016), there have been many reports of depopulation occurring across entire cities, including the emergence of so-called "spongy" urban areas with many sporadically located vacant lots and empty houses. Therefore, it is important to understand how the spatial composition of population density within cities has evolved in order to evaluate urban policies implemented to the present date and then consider future directions for urban policies.

The spatial-cycle model (SCM) presented by Klaassen et al. (1981) is a well-known tool for modeling the stages of population growth and change within a city. According to this model, a city is categorized into two regions: the city center and the suburbs. Then, by focusing on the rate of population change in both regions, the city's current status can be categorized into one of four stages: (1) urbanization, (2) suburbanization, (3) disurbanization, and (4) re-urbanization. It is assumed that all cities go through each of these stages in order.

The first stage, urbanization, is that in which growth in the city center drives the growth of the city as a whole. The second stage, suburbanization, is when the city experiences growth in the population of the suburbs. In the third stage, disurbanization, the population in the city center declines, eventually resulting in the decline of the city as a whole. In the fourth stage, reurbanization, the suburbs continue to decline but the city center regains its population.

The urbanization stage corresponds to a situation that was widespread in European cities during the post-war years of the 1950s. This is the stage in which the development of industry in cities caused rural-urban migration patterns toward cities and away from villages. This was followed in the 1960s by a suburbanization stage that was generally caused by the growth of automobile ownership and transportation networks connecting city centers and the suburbs. This stage was further stimulated by the emergence of a new suburban middle class based on rising income levels (Ishikawa 2008; Sakiyama 1981). The next stage, disurbanization, came to be identified as a prominent counter-urbanization trend in the 1970s. It has been pointed out that this phenomenon has taken place against the backdrop of inner-city decay surrounding the central business districts as well as changes in employment structure resulting from post-war industrialization (Kato 1985; Narita 1988).

The reurbanization stage was put forward as a hypothetical fourth stage when the SCM was first proposed. But in the 1980s, some cities actually did experience reurbanization, mainly in Western Europe. This phenomenon can be attributed to the redevelopment of inner cities and other urban centers through a process of gentrification, which improved the urban environment and increased the central population (Narita 1995; Kabisch and Haase 2011; Pacione 2001). Accordingly, it took roughly 30–40 years for cities in post-war Europe to sequentially go through each of the above four stages.

Furthermore, Geyer and Kontuly (1993) proposed a theory of 'differential urbanization,' in which larger cities go through the stages of development from urbanization to reurbanization before smaller cities, and they redefined SCM from an urban system perspective, rather than as a model that merely describes the urban structure of a single city.

Several prior studies have sought to grasp the urban structure of Japan based on SCM. For example, Ikegawa (2001) and Kawashima (2001) used the concept of the ROXY index to identify cyclical trends in the concentration and dispersion tendencies of urban structures. Kim et al. (2007) broadly affirmed the cyclical nature of SCM based on an analysis of urban structures built over the 30-year period up to 2000, and they noted a strong trend toward spatial dispersion of urban populations during this period. However, the above studies did not assess the situation beyond the year 2000. Although a more recent analysis was performed by the Kyushu Economic Research Center (KER 2015), this only analyzed population changes over a five-year period (2005–2010) and only considered cities in the Kyushu region.

On the other hand, there are newer studies suggesting that the transitions of urban structures do not actually follow the sequence assumed by SCM. For example, Wolff (2018) conducted an empirical study to grasp the urban demographics of 36 European countries from 1990 to 2010 and found that, in a certain number of cases, the urban structures had transitioned contrary to the expectations of SCM. He pointed out that the order in which urban development phases are experienced varies from one city to the next. Based on his findings regarding the population distribution dynamics in the centers and suburbs of Japanese cities, Yamagami (2006) reported that in the five-year period from 1995, larger cities experienced the same suburban population decline and spatial concentration trends that smaller and medium-sized cities were already experiencing. Furthermore, in the period after 2000, he predicted a trend contrary to the theory of differential urbanization, that is, where larger cities experience specific demographics *after* they are experienced by small and medium-sized cities.

Although the cyclical behavior assumed by SCM is open to question, its approach of dividing cities into central and suburban areas as a way of grasping population distribution is nevertheless a useful way of discussing the changes in urban structure. Therefore, the purpose of this paper is to apply SCM to 109 Japanese cities over the period of 35 years leading up to the most recently available Census data (1980–2015). We use recent data to judge whether the cycle assumed by SCM can be verified, examine whether it can support the differential urbanization theory, and grasp the changes in urban structure that Japanese cities have experienced during the era of national population decline.

4.2 Research Method

4.2.1 Spatial Unit of Analysis

We adopt Urban Employment Areas (UEAs), the travel-to-work areas defined in Kanemoto and Tokuoka (2002), as the spatial unit of cities. The UEA scheme basically identifies a municipality with above-threshold population within its densely inhabited districts as a central municipality, and a municipality from which more than 10% of its working population commute to a central municipality as a suburban municipality. A UEA can have multiple central municipalities, so a polycentric urban structure is supported to some extent. For example, the Tokyo UEA includes Tokyo Wards, Kawasaki City, Yokohama City, Chiba City, and four other municipalities as its central municipalities. The UEAs are updated every five years based on the national census, but in this study, we fix the units and use the data for the year 2000, which lies in the middle of our study period of 1980–2015.

4.2.2 Application of Spatial-Cycle Model

The basic idea of SCM is to determine whether the population of the city centers and their suburbs are growing or declining, and to classify the state of these cities as urbanization, suburbanization, disurbanization, or reurbanization based on the results. This classification itself can thus be done without assuming anything about the order in which these stages occur. In this chapter, as shown in Fig. 4.1, each city is plotted on a two-dimensional graph with population change rates in the center on the horizontal axis and population change rates in the suburbs on the vertical axis. Consequently, it is now possible to classify which stage of urban structure a city is currently experiencing based on its plotted location.

In this study, the population change rates in the center are measured based on the rate of population change in the central municipalities of the UEAs, and the population change rates in the suburbs are measured based on the rate of population change in the suburban municipalities of the UEA. We used the Census results from 1980 to 2015 as our source of data for the municipal population.[1] We then calculated the rate of population change for each 10-year or 5-year period from 1980 to 2015 based on the areas of central and suburban municipalities in the urban employment area as of the year 2000.

As shown in Fig. 4.1, when cities are plotted on a plane at coordinates corresponding to the rates of population change in the city center and suburbs, they can

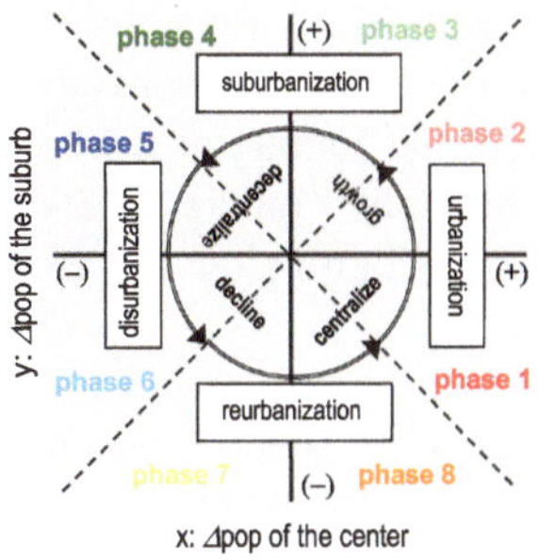

Phases	Stages	Δpop per 10 years		Pop. size	Pop. distribution
		center	suburb		
0	(no change)	± 2%	± 2%	(no change)	(no change)
1	urbanization	+ +	−	growth	centralize
2	urbanization	+ +	+	growth	centralize
3	suburbanization	+	+ +	growth	decentralize
4	suburbanization	−	+ +	growth	decentralize
5	disurbanization	− −	+	decline	decentralize
6	disurbanization	− −	−	decline	decentralize
7	reurbanization	−	− −	decline	centralize
8	reurbanization	+	− −	decline	centralize

Fig. 4.1 Spatial-cycle model: concepts and definitions. Adopted from Klaassen et al. (1981)

be classified into eight phases (phase 1 through phase 8). The previously mentioned works of Wolff (2018) and KER (2015) also use these eight categories (or aggregates thereof) to categorize the stages of urban structure, and we basically follow the same approach. However, if the central and suburban municipalities of the urban employment area both have population change rates with small absolute values, i.e., if the city is plotted near the origin in Fig. 4.1, then the urban structure will hardly change at all no matter which phase the city is categorized into. Therefore, for convenience, such cities are recategorized as phase 0 (no change in urban structure) if the rate of population change in the central and suburban municipalities is within $\pm$ 2% over 10 years or $\pm$ 1% over 5 years. If the threshold is set within $\pm$ 2% over 10 years ($\pm$ 1% over 5 years), approximately 10% of the cities will be classified as phase 0 from each year to the next. However, the setting of this threshold does not significantly affect the subsequent results and conclusions.[2]

Next, we explain the perspective used to interpret what kind of urban structural changes a city is undergoing according to these categorized phases. First, Fig. 4.1 shows the correspondence between the phase categories and the name of each stage of SCM. For example, phase 1 and phase 2 correspond to the so-called urbanization stage in SCM. It should be noted that the names of SCM stages, such as "disurbanization" and "reurbanization" (along with the term "stage" itself), presuppose a sequence that follows "urbanization" and "suburbanization." Although this terminology is not strictly appropriate if the city does not transition in the sequence envisioned by SCM, we nevertheless use these terms in this chapter as labels to categorize the changes in urban structure.

In the cities categorized into phases 1, 2, 7, and 8, which fall below the 45° line (the line $y = x$) in Fig. 4.1, the population growth rate of the central municipalities is greater than that of the suburban municipalities. In other words, these are interpreted as cities that exhibit a centralization trend, with a relative concentration of population in the central municipalities is occurring. Conversely, the cities above the 45° line (phases 3, 4, 5, and 6) are interpreted as decentralizing cities with populations that are dispersing toward the suburban municipalities. Furthermore, cities categorized into phases 1, 2, 3, and 4 that lie above the 135° line (the line $y = -x$) in Fig. 4.1 are interpreted as having overall populations that are increasing. On the other hand, cities categorized into phases 5, 6, 7, and 8, which are below the line, are interpreted as having overall populations that are decreasing.[3] Therefore, in general, more cities are categorized above the 135° line when the population of the country as a whole is increasing. On the other hand, if the population of the entire country enters a declining phase, the distribution of cities will generally transition below the 135° line.

Furthermore, as pointed out by Geyer and Kontuly (1993), the urban structure that people experience may differ depending on the scale of the city. Therefore, in this paper, the scale of a city is interpreted as being represented by its population, and the cities we analyze are divided into four size classes according to the population of the functional urban region as of the year 2000: "million cities" (at least 1 million people), "1/2 million cities" (at least 500,000 people), "1/4 million cities" (at least 250,000 people), and "smaller cities" (less than 250,000 people). We cross-tabulated the number of cities categorized into each phase against these four city-size classes. In the following, a city of 1 million people is described as a large city, a city of 500,000 as a medium-sized city, and so on.

4.3   Results

4.3.1   Results of Applying Spatial-Cycle Model

Of the 113 major UEAs defined in 2000, 109 UEAs for which central and suburban municipalities could be defined were plotted on planar coordinates with the rate of population change in central municipalities on the horizontal axis and the rate of population change in suburban municipalities on the vertical axis as shown in Fig. 4.2. Table 4.1 shows the results of categorizing the urban development phases of these 109 UEAs, and Fig. 4.3 shows this data plotted on a map.

The results show that in the 1980s, many of the "million cities" experienced suburbanization in phases 3 and 4. Meanwhile, many of the medium-sized ("1/2 million" and "1/4 million") cities experienced urbanization or suburbanization. However, some of the smaller cities in the peripheral areas of Japan, such as Hokkaido and Kyushu, were categorized as disurbanization or reurbanization (phases 5–8).

Next, we can see that in the 1990s, a larger proportion of cities experienced suburbanization, not only among the "million cities" but also among the "1/2 million" and "1/4 million cities." Most of the cities categorized in the suburbanization phase belonged to phase 3, where the population of the central municipalities is increasing. However, the Osaka metropolitan area belonged to phase 4 because the population of its central municipalities was declining. Furthermore, of the medium-sized cities in the Chugoku and Shikoku regions, it can be seen that cities that were in the urbanization category in the previous decade tended to continue an urbanization trend instead of transitioning to suburbanization. In the Tohoku region and in other

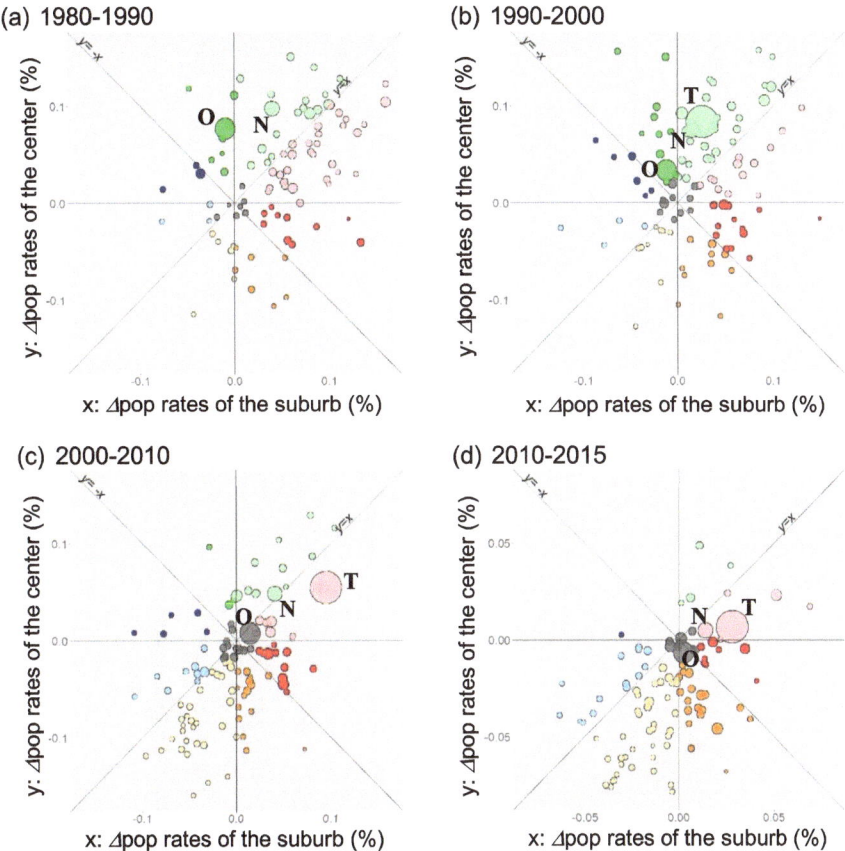

Fig. 4.2 Population change rates in the suburbs and the centers of Japanese cities 1980–2015. The symbol size corresponds to the city population and the color to the phases as defined in Fig. 4.1. In the graphs, T, O, and N are the three major cities of Tokyo, Osaka, and Nagoya, respectively. Tokyo in 1980–1990 is located outside of the figure, to the upper right, and is in phase 3

peripheral areas of Japan, some cities have been consistently categorized into the reurbanization stage since 1980.

In the 2000s, there were more cases of large cities being categorized into the urbanization stage. These trends in large cities correspond to the phenomenon known as "the return to the city center," or *toshin-kaiki*, in which population of the city center increased. This happened without going through the disurbanization stage in Japanese and thus distinguished from reurbanization. Among the larger cities, Nagoya and many of its surrounding cities continued to experience suburbanization as they did prior to 2000. Furthermore, among the smaller cities, the number of cities in the reurbanization category is increasing. Finally, in the five years from 2010, with the exception of some large-scale cities, most cities were in the reurbanization category. In the case of cities in Hokkaido, even where the population has declined

Table 4.1 Number of cities in SCM phases for each period, by city-size classes

	Million cities				1/2 million cities				1/4 million cities				Smaller cities				Total			
	'80–90	'90–00	'00–10	'10–05	'80–90	'90–00	'00–10	'10–05	'80–90	'90–00	'00–10	'10–05	'80–90	'90–00	'00–10	'10–05	'80–90	'90–00	'00–10	'10–05
Phase 0		1	1	3		2	6	3	3	6	6	2	4	1			7	10	13	8
Phase 1		1	3	2		3	6	1	4	3	1	2	4	6	1	4	10	13	11	9
Phase 2	4		4	3	14	6	2		13	6	2	2	6	1	2	1	37	13	10	6
Phase 3	5	9	3	1	8	9	3	1	8	10	2		2	4	3	2	23	32	11	4
Phase 4	2	1			3	5	1		1	4	2		5	1			11	11	3	
Phase 5	1					1			2	1	2			4	3	1	3	6	5	1
Phase 6			1				2	3	1	2	4	7	3	2	2	5	4	4	9	15
Phase 7				1			3	10	2	1	13	15	4	8	16	20	6	9	32	46
Phase 8				2		1	4	9	2	3	4	8	6	7	7	1	8	11	15	20
Growth	11	11	10	6	27	23	12	2	26	23	7	4	17	12	6	7	81	69	35	19
Decline	1		1	3		2	9	22	7	7	23	30	13	21	28	27	21	30	61	82
Centralize	4	1	7	8	16	10	15	20	21	13	20	27	20	22	26	26	61	46	68	81
Decentralize	8	10	4	1	11	15	6	4	12	17	10	7	10	11	8	8	41	53	28	20
Total	12	12	12	12	27	27	27	27	36	36	36	36	34	34	34	34	109	109	109	109

Note Growth/decline and centralize/decentralize are reclassifications of the phases. See Fig. 4.1 for definitions of the phases

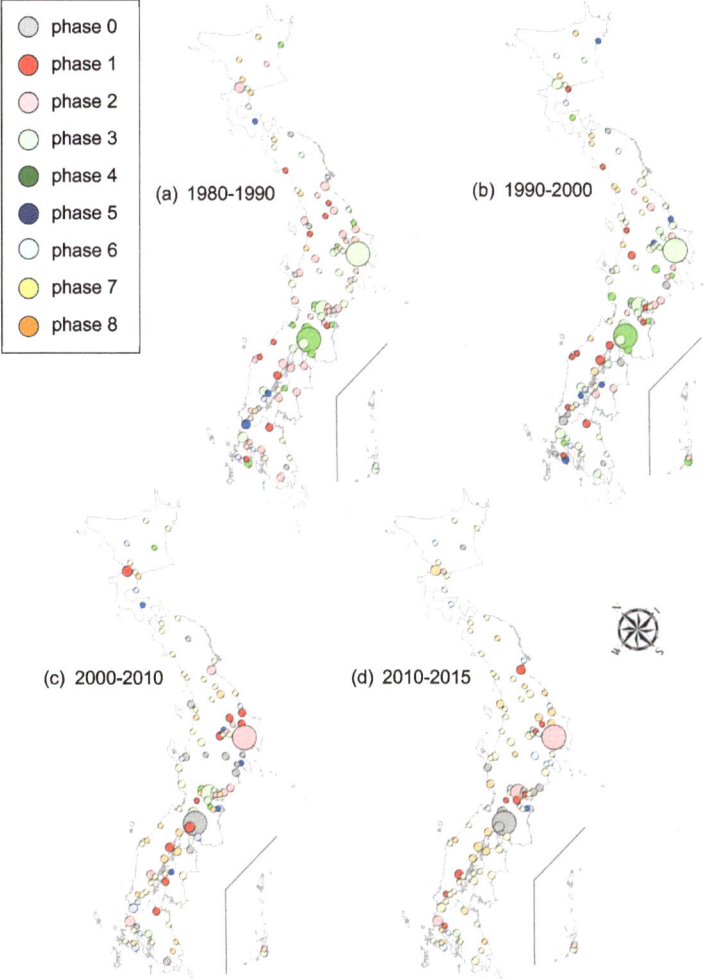

Fig. 4.3 Spatial distribution of Japanese cities in each phase. The color of the symbol represents the phases and the size of the symbol corresponds to the four city-size classes

in both the city center and the suburbs, the percentage of cities that are in phase 6 rather than phase 7 (i.e., cities that are declining in population while showing a trend toward decentralization) has been consistently higher than the percentage expected nationwide since 1980, and markedly so since 2000.

To summarize the above trends since 2000, most Japanese cities shifted below the 45° line (the line $y = x$), and this trend has been particularly strong in cities where the overall population has been declining. In other words, our analysis of cities based on this division between central municipalities and suburban municipalities as of 2000 suggests that many cities have been exhibiting a tendency toward "centralization."

4.4 Discussion

4.4.1 Sequential Nature of the Spatial-Cycle Model

SCM assumes that urban structures evolve according to the sequence of urbanization followed by suburbanization, disurbanization, reurbanization, and then back to urbanization. In this section, we examine whether such a cycle existed in Japanese cities from 1980 to 2015. For this section, we follow the methodology of Wolff (2018), who undertook a similar discussion of SCM sequence in European cities. Accordingly, in Fig. 4.1, a city might belong to phase 1 for a 10-year period, such as 1980 to 1990, and then remain in phase 1 for the next 10-year period, here from 1990 to 2000. In such cases where a city belongs to a particular phase in one decade (or five-year period) and then belongs to the same phase in the next decade, this transition pattern is classified as "same phase." Furthermore, if a city is classified in a particular phase for one 10-year (or 5-year) period and then shifts by one phase in the direction of the arrow in Fig. 4.1 in the following 10-year period, it is regarded as being "in order." If it shifts by two or three phases in the direction of this arrow, it is regarded as being "in skipping order." Similarly, if a city shifts by one phase or by two or three phases in the opposite direction to that of the arrow, it is regarded as being "in reverse order" or "in skipping reverse order," respectively. In addition, if the phase transitions from its original position to a point-symmetrical position with the origin as the center of symmetry (for example, from phase 1 to phase 5), it is categorized as being "point symmetrical," and transitions from other phases to phase 0 and vice versa are categorized as transitions "to and from phase 0."

Table 4.2 shows the results of examining the sequential nature of SCM based on the above reclassification method. About half of the 109 cities underwent urban structural transitions during each year. Of these, the number of transitions in reverse order is comparable to the number of transitions in order. Therefore, the recent urban structure of Japanese cities has not necessarily transitioned in the order envisioned by SCM. On closer inspection, in-order transitions were more dominant until 2000, while reverse-order transitions became more dominant after 2000.

4.4.2 Discussion on Centralization of Urban Structures

In order to gain an understanding of whether urban population distributions tend to become concentrated or diffused, we reclassified the results of dividing the cities into eight phases (excluding phase 0) by grouping them according to whether they were undergoing centralization or decentralization (Table 4.1). The results show that more cities decentralized between 1990 and 2000, and this was especially true for the larger classes of city. In contrast, few cities became decentralized after 2000, and this was true for all city classes. This shows that most of Japan's cities, regardless of size, have experienced an increasing concentration of population in their city centers since

Table 4.2 Transition patterns between phases

Transition patterns	Starting/ending periods		
	'80–90/'90–00	'90–00/'00–10	'00–10/'10–15
1. Same phase	51	31	46
2. In order	27	11	12
3. In skipping order	5	7	3
4. In reverse order	13	26	26
5. In skipping reverse order	2	13	3
6. Point symmetrical	0	2	0
7. To or from phase 0	11	19	19
Total	109	109	109

Note The entry for starting/ending period '80–90/'90–00, for example, indicates the transition in phases between the periods 1980–90 and 1990–2000. See text for definitions of transition patterns

2000. Although this result differs from the trend reported by Kim et al. (2007), this difference can be attributed to the fact that they only analyzed the period up to 2000 and examined city centers over a narrower scope than in this study. Therefore, when discussing the tendencies of population concentration and dispersion by dividing cities into central and suburban areas, the results may differ depending on how the central and suburban areas are separated. In Kanda et al. (2020), gridded population data were used to study the population-weighted average of the distance to the city center from each grid cell, and it was found that this average value had declined for all city-size classes since 2000.

4.4.3 Differences in Urban Structural Transition Processes by City Size

Looking at the results of Fig. 4.2 once again, many Japanese cities have moved from the first quadrant (phases 2 and 3) to the fourth quadrant (especially phase 7) over time. This trend can also be seen in Table 4.1. It could be said that the trend toward centralization of urban structures alongside declining urban population arose from relatively small cities with a population of 250,000 or less, or from cities at the periphery of the country, and it is now spreading to medium-sized and larger cities.

One factor that determined the decentralization of urban structures is the increased pressure to develop residential areas in the suburbs during the bubble economy era in the late 1980s. In other words, the increased demand for office space in urban centers during the bubble promoted the development of outlying suburbs around large cities. This situation can be viewed as a phenomenon in which the spatial extent of cities expands in response to urban economic growth. Conversely, the centralization of the urban structure can be attributed to redevelopment efforts, including the construction of high-rise residential buildings, in urban centers following the collapse of the bubble economy. This can be viewed as a phenomenon in which city centers became taller and denser in response to the increased urban population. However, the pressure to develop suburban housing during the bubble economy era was greater in large cities, and although there are some reports of population recovery due to the redevelopment of urban centers in regional cities (Nakamura 2016), the scale of the housing supply has remained greater in larger cities. In other words, the redevelopment of city centers is not a sufficient explanation for why small cities experienced concentration of urban structures before large cities. Other possible factors are presented below.

First, in rural areas, population declines are greater than in cities due to people seeking employment and educational opportunities elsewhere in addition to higher death rates due to a higher proportion of the aged population (MAFF 2018). Therefore, the centralization of urban structures seemed to occur earlier in smaller cities that have more rural areas within their suburbs than in larger cities. Previous studies that analyzed the population of Japanese municipalities have shown that municipalities with smaller populations tend to have higher population decline rates (Esaki 2016; Takano 2010). This can be confirmed from the growth/decline rates obtained by reclassifying the phases of cities as shown in Table 4.1. Furthermore, it has been reported that municipalities with larger rates of population decline also have larger rates of decline in the number of business offices (SME 2016). In light of the above observations and considering the fact that relatively many municipalities with small populations are distributed in the suburbs of small cities, it can be inferred that the number of business offices has decreased significantly in the suburban municipalities of small cities. By definition, a UEA adopts a threshold of 10% for the commuting rate, so the suburban municipalities actually have many residents who commute within their own municipalities in addition to those who commute to the central municipalities. Accordingly, a reduction in employment opportunities in the suburban municipalities of small and medium-sized cities affects the demographics of these municipalities, and as a result, the concentration of the urban structure tends to start in small cities.

In regional cities, the constituent ratio of secondary industries, such as manufacturing and construction, is higher than in metropolitan areas (SME 2016), and the number of people working in these industries is on a downward trend over the long term (MHLW 2013). On the other hand, Japan has experienced changes in its industrial structure, with the rise of urban-based service industries replacing secondary industries. Suzuki (1992) pointed out that while the development of service industries has increased employment opportunities in regional core cities such as prefectural capitals, the growth potential of tertiary industries has been relatively weak in the surrounding smaller cities, resulting in the creation of local "growth poles" for the core cities in each region.

As described above, in Japan, the population is declining in smaller cities, and as a result, the urban structure of these smaller cities is showing a tendency to become more concentrated. This is contrary to the differential urbanization theory of Geyer and Kontuly (1993), which says that large cities experience changes in urban structure before small cities, but consistent with the prediction of Yamagami (2006) that large cities will experience the demographic trends that were experienced by small and medium-sized cities after 2000. In the future, if the national population decline is not reversed, by immigration policies and the like, then the trend of declining population will also apply to large cities. Indeed, larger cities in Japan will probably experience the changes that smaller cities have already experienced.

4.5 Conclusion

In this study, we discussed the evolution of urban structure in Japan's urban employment areas (UAEs) based on the spatial-cycle model (SCM). As a result, we found that since the year 2000 Japanese cities have changed more frequently in the opposite direction to the sequence assumed by SCM. This result was obtained after removing the possibility that the population change rates might fluctuate slightly between positive and negative in the vicinity of 0%, thus disrupting the phase sequence.

However, the urban structures of Japanese cities are not changing in a chaotic manner. In other words, although the decline of city centers and the decentralization of cities due to suburban development are still cited as urban policy problems today, few cities continued to spatially decentralize their population distribution between 1980 and 2015. On the contrary, our results show that since 2000, the population distribution of cities as a whole has tended to become more concentrated.

In Japan, changes in urban structure experienced by smaller cities and peripheral areas of the country are experienced somewhat later by larger cities and central areas of the country, which is contrary to the predictions of differential urbanization theory. As the growth of cities as a whole is showing signs of restraint against the background of ongoing nationwide changes in the industrial structure and a declining population, this can be regarded as a phase of shrinkage of suburban areas, not only in provincial cities but also gradually in the larger cities.

SCM is a model of the 'stages of urban development.' When the national population declines and the cities shrink, the 'stages of urban retrogression' ought to be in the reverse order, with the effect 'trickling up' from the smaller cities to the larger ones. Consequently, SCM and the differential urbanization theory were right, after all.

Acknowledgments The original work in this chapter is based on a master's thesis submitted to the Graduate School of Science, Tohoku University, by the first author. All findings and conclusions expressed in this chapter are those of the authors and do not necessarily reflect the views of Japan's Ministry of Land, Infrastructure, Transport and Tourism.

Notes

(1) In the National Census, population data from 2000 onward can be aggregated by municipality as of 2000 thanks to the availability of data aggregated by former municipalities prior to the Great

Heisei Amalgamation of Japanese municipalities. The method used to match municipalities at arbitrary times between 1980 and 2015 with municipalities as of 2000 was based on the work of Kirimura et al. (2011).

(2) The classification criteria for phase 0 are that the population change rate in central municipalities and suburban municipalities remains within ± 1% over 10 years (or within ± 0.5% over 5 years) or that separate urban structures are reclassified by changing these criteria to within ± 3% over 10 years (or within ± 1.5% over 5 years). We compared these classification results with the results obtained using the classification criterion described in this report (population change rate within ± 2% over 10 years or within ± 1% over 5 years) and found no significant difference.

(3) If the population of the central municipalities exactly matched the population of the suburban municipalities, the 135° line would strictly be the boundary line of population increase or decrease for the entire city. But in general, the populations of the central municipalities and suburban municipalities are different. However, even when they do differ, the 135° line can generally be interpreted as the boundary line.

References[1]

Esaki Y (2016) Nihon no chiho toshi ni okeru jinko henka (Population trends of cities in rural regions of Japan). Chigaku Zasshi (Journal of Geography) 125(4): 443–456. https://doi.org/10.5026/jgeography.125.443 (J)

Geyer HS, Kontuly T (1993) A theoretical foundation for the concept of differential urbanization. International Regional Science Review, 15, 157–177. https://doi.org/10.1177/016001769301500202

IPSS (2018) Jinko tokei shiryosyu 2018 (Demographic data collection 2018), Jinko Mondai Kenkyu Shiryo (Population Studies Research Materials) 338:7–40. (J)

Ikegawa S (2001) Wagakuni no toshi saikuru to toshi seibi no hoko: ROXY shihyo ni yoru sengo yaku 50 nenkan no bunseki (Urban cycle and direction of urban development in Japan). Sogo Kenkyu (Comprehensive Research) 20: 5–26. (J)

Ishikawa Y (2008) Kogai kara mita tosiken kukan (Urban space from the suburban perspective). Kaiseisha Press, Otsu. (J)

KER (2015) 2015 nenban kyushu keizai hakusyo (2015 Kyushu economic white paper). Kyushu Keizai Kenkyujo (Kyushu Economic Research Center), Kagoshima. (J)

Kabisch N, Haase D (2011) Diversifying European agglomerations: Evidence of urban population trends for the 21st century. Population, Space and Place, 17(3), 236–253. https://doi.org/10.1002/psp.600

Kamezawa H (2010) Chushin shigaichi kasseika kihon keikaku no jisshi jokyo to kongo no kadai (Implementation of the basic plan for the revitalization of central city areas and future issues). Rippo to Chosa (Legislation and Research) 40:2–7. (J)

Kanda H, Isoda Y, Nakaya T (2020) Jinko gensho kyokumen ni okeru nihon no toshi kozo no hensen (The transition of urban structure in Japan in the phase of decreasing population). Kikan Chirigaku (Quarterly Journal of Geography) 72:91–106. https://doi.org/10.5190/tga.72.2_91 (J)

Kanemoto Y, Tokuoka K (2002) Nihon no toshiken settei kijun (Proposal for the standards of metropolitan areas of Japan). Oyo Chiiki Kenkyu (Journal of Applied Regional Science) 7: 1–15. (J)

Kato Y (1985) Toshi no kozo henka to kadai (Urban structural changes and challenges). In: Sakamoto H, Hamatani M (eds) Saikin no chirigaku (Recent Geography), Taimeido, Tokyo. p 185–194. (J)

---

[1] (J): written in Japanese

Kawashima T (2001) Toshi saikuru to jutaku seisaku (Urban cycle and housing policy). Jutaku Tochi Keizai (Housing Land Economy) 40:2–7. (J)

Kim C, Onishi T, Suga M (2007) Jinko gensho to toshi kozo no henyo ni kansuru kenkyu (Study on depopulation and transformation of urban structure). Toshi Keikaku Ronbunshu (Journal of the City Planning Institute of Japan) 43(3): 835–540. (J)

Kirimura T, Nakaya T, Yano K (2011) Shikuchoson no kuiki ni kansuru jikukantekina chiri joho deta besu no kaihatsu (Building a spatio-temporal GIS database about boundaries of municipalities). GIS Riron to Oyo (Theory and Applications of GIS) 19(2): 83–92. https://doi.org/10.5638/thagis.19.139 (J)

Klaassen LH, Bourdrez JA, Volmuller J (1981) Transport and reurbanisation. Gower, Aldershot.

MAFF (2018) Heisei 30 nenndo shokuryo, nogyo, noson hakusho (2018 White paper on food, agriculture, and rural areas). Ministry of Agriculture, Forestry and Fisheries, Tokyo. (J)

MHLW (2013) Heisei 25 nenban rodo keizai hakusho (2013 white paper on labor economy). Ministry of Health, Labour and Welfare, Tokyo. (J)

MLIT (2018) Ricchi tekiseika keikaku sakusei no torikumi jokyo (Status of efforts to create the Site Optimization Plans). Ministry of Land, Infrastructure, Transport and Tourism, Tokyo. http://www.mlit.go.jp/toshi/city_plan/toshi_city_plan_fr_000051.html. Accessed 30 Oct 2018 (J)

Nakamura T (2016) Chiho toshi ni okeru chushin shigaichi no jinko kaiki no jittai (Recent repopulation of local cites' central areas). Toshi Keikaku Ronbunshu (Journal of the City Planning Institute of Japan) 51(2): 159–166. https://doi.org/10.11361/journalcpij.51.159 (J)

Narita K (1988) Daitoshi suitai chiku no saisei (Revitalization of declining metropolitan areas). Taimeido, Tokyo. (J)

Narita K (1995) Tenkanki no toshi to toshiken (Cities and metropolitan areas in transition). Chijin Shobo, Kyoro. (J)

Nikkei (2018) Genkai toshi (Marginal cities). Nihon Keizai Shimbun, 21 April 2018, Morning Edition. (J)

Pacione M (2001) Urban geography: A global perspective third edition, Routledge, Abingdon.

SME (2016) Shokibo kigyo hakusho (White paper on small and medium enterprises in Japan). The Small and Medium Enterprise Agency, Tokyo. (J)

Sakiyama K (1981) Toshika to daitoshi mondai (Urbanization and metropolitan issues). In: Yoshioka K, Sakiyama K (eds) Daitoshi no suitai to saisei (Decline and revival of large cities). University of Tokyo Press, Tokyo, p 3–28. (J)

Suzuki S (1992) Transformation of industrial structure and local cities (Local city under industrial structure change). Sangyo Gakkai Kenkyu Nenpo (Annual Report of the Japan Society of Industrial Science and Technology) 8: 17–36. https://doi.org/10.11444/sisj1986.1993.17 (J)

TMG (2016) Tomin fasuto de tsukuru "atarashii tokyo" (A "New Tokyo" created by the people first). Tokyo Metropolitan Government, Tokyo. (J)

Takano T (2010) Jinko gensho jidai o mukaeru tohoku chiho no toshi shisutemu no doko (Trends in urban systems in the Tohoku Region in an era of declining population). Tohoku Bunka Kenkyujo Kiyo (Tohoku Culture Research Institute) 42: 18–34. (J)

Ujihara T, Abe H, Murata N, Washio N (2016) Chiho toshi ni okeru toshi sponjika no jisshoteki kenkyu (Reality of "spongy urban area" in local city). Doboku Gakkai Ronbunshu (Journal of JSCE D3) 72(1): 62–72. https://doi.org/10.2208/jscejipm.72.62 (J)

Wolff M (2018) Understanding the role of centralization processes for cities: Evidence from a spatial perspective of urban Europe 1990–2010. *Cities*, 75, 20–29. https://doi.org/10.1016/j.cities.2017.01.009

Yamagami T (2006) Nihon ni okeru toshiken no jinko kibo to tosiken nai no jinko bunpu no hendo to kankei (The relationship between metropolitan size and the population redistribution pattern within metropolitan areas in Japan). Jimbun Chiri (Human Geography) 58(1): 56–72. https://doi.org/10.4200/jjhg.58.1_56 (J)

Chapter 5
Significance and Possibilities of the New Concept of "Relationship Population" in Japan's Population Decline Society

Hirokazu Sakuno

Abstract This chapter examines the new concept of "relationship population," which originated in Japan after 2016, and investigates it by presenting a new view on the significance of this concept. Generally, relationship population has been regarded as the third population concept, situated between exchange population and residential population. The author does not deny these interesting characteristic of relationship population, but he also thinks that this new population concept should not be grasped only as a fixed stage between those two previous population concepts. Instead, it should be regarded as a concept that includes various rich meanings, particularly between urban centers and agricultural/fishing villages in the new era of depopulation. Therefore, this paper attempts to clarify the diverse elements involved in this new concept. Specifically, the author categorizes the significance types of relationship population into four specific groups based on the different relationships they use to connect urban and rural areas. Then, the characteristics of each type are discussed in detail. In a previous discourse on relationship population, more attention was paid to the aspects associated with the regional-support-oriented and regional-contribution-oriented types. In addition, the importance of the slow-life-oriented and non-residential-area-maintenance types also needs to be recognized.

Keywords Population · Depopulation society · Relationship population · Sustainability · Community development · Local innovation

5.1 Introduction

In Japan, since the 1990s, the continued flow of the population from rural areas to urban areas and the aging of residents have led to the decline of the population in rural areas. However, the decline in the population of Japan as a whole, starting around 2005, has led to the movement of population from urban to rural areas,

H. Sakuno (✉)
Faculty of Education, Shimane University, Matsue, Shimane, Japan
e-mail: hsakuno@edu.shimane-u.ac.jp

73

which has become a social phenomenon. First, in the early 2000s, the phenomenon called *furusato kaiki* (return to hometown) was observed, in which retired people, mainly baby boomers, returned to their hometowns. Then, when the Great East Japan Earthquake and the nuclear disaster occurred in 2011, they triggered an increase in the number of young people in their 20 and 30s who became interested in or have actually moved to rural areas.

Although this phenomenon is partly attributable to the changes in people's values as a result of the declining population in society, the impact of government policies is also considered significant. For example, since the Community-Reactivating Cooperator Squad (CRCS) program started in FY2009, more than 6000 urban residents have volunteered to engage in rural activities.[1] In FY2014, the Regional Revitalization Act was enacted. Consequently, along with the National Comprehensive Strategies, the Municipal Version of the Comprehensive Strategies and the Municipal Population Vision were formulated for all municipalities nationwide. This resulted in wide recognition of the importance of promoting the flow of population from urban to rural areas as a national project. This has in turn led to an unprecedented increase in the movement of people from urban to rural areas as well as a stronger inclination among urban residents to move to those rural areas—a phenomenon that has come to be known as "migration to the countryside."

However, although this migration to the countryside has become widespread, the outflow of the population from rural to urban areas has not subsided. In other words, despite this phenomenon of people migrating to the countryside, the population in rural areas has not recovered, nor has the population decline abated. This has led to a growing sense of entrapment and resignation among the people in some rural areas.

Recently, local governments have come to recognize the difficulty of increasing or maintaining the existing residential population, and thus they have instead decided to increase the exchange of population as a policy issue. However, it has also become gradually clear that it is difficult to sustain regional populations even if the exchange of population increases. Consequently, the concept of "relationship population" has been proposed as a solution to this deadlock.

This study aims to redefine the concept of relationship population and examine its significance from the following perspectives.

First, we consider the standpoint that each region should pursue regional sustainability as the appropriate direction, i.e., in terms of how the relationship population is related to community development[2] in pursing sustainability. In the past, each region was expected to achieve economic competitiveness and, therefore, prioritized increasing its population, attracting enterprises, and enhancing convenience in transportation and logistics. While these strategies are also important for the future, Japan as a whole has become a "depopulation society," in which only so much can be done to implement public policies based on increasing the population and enhancing local industries. In the future, regions should work toward enhancing their distinctiveness and sustaining their economy in a way that best suits their characteristics and needs. The author believes that one of the ways to achieve this is to establish relationship population, and in this paper, he shows the effectiveness of relationship population in creating regional sustainability.

Second, we examine the standpoint that the autonomy of residents is essential for achieving regional sustainability. There are two types of autonomy, namely group autonomy and resident autonomy. As a result of the decline in rural areas and calls for regional revitalization, national and local governments have proposed regional revitalization policies one after another. However, the more focus is given to policies, the more likely is the emergence of a structure in which the government takes the initiative while residents can only respond obediently. Therefore, in order to promote the autonomy of residents, it is necessary to form a Regional Management Organization[3] in each region. However, a frequent problem here is the lack of players and successors. In most regions, the prospect of maintaining the residential population is nil, so there are expectations on the potential of relationship population. Specifically, the spread of relationship population is expected to underpin the autonomy of residents and build sustainable regions.

Third, we look at the standpoint of local innovation. In order to sustain the region, it is important to maintain the minimum manageable population in the community. However, not all issues can be addressed simply by ensuring a certain level of population. In particular, in rural areas, residents have no choice but to keep up with the rapidly changing lifestyles and production modes of modern society, while also maintaining the traditions of the past. Therefore, it is necessary to acquire new perspectives and ideas without getting encumbered by established notions. Doing this requires the proactive involvement of stakeholders from outside the region, since there is a limit to what locals can do. In the past, this role has been played by migrants, such as the so-called I-turn migrants; however, the relationship population will likely play an additional major role. Relationship population is expected to bring flexibility to the formalized consensus-building system and effect transformation in the regions.

From these three standpoints, this paper aims to clarify the concept of relationship population and to realign existing policies and demographic phenomena from the perspective of the relationship population. Accordingly, we examine how sustainability can be ensured in each region.

5.2 Emergence and Concept of Relationship Population

5.2.1 Emergence of the Relationship Population Concept

Relationship population is a new concept that spread between 2016 and 2017. It is believed to have originated from the following three publications released within a year, after which the term "relationship population" rapidly spread throughout society.

The first is *Mixing urban and peripheral areas* by Takahashi (2016). Takahashi claimed that even in urban areas, which are thought to face fewer local problems than rural areas, the "hardships" of residents have increased, similar to what is happening in marginal cities. The options for U-turn and I-turn migration[4] make it possible to overcome this situation, but the hurdle is too high in risking everything to make the move to rural areas. Meanwhile, many of the volunteers who came to assist the recovery from the Great East Japan Earthquake, for example, continued to stay involved in the rural areas even though they did not move permanently into the area. In other words, thus far, the premise of involvement with a certain rural area was residence, and the choice was to either "reside or not." Takahashi, however, showed that there are various ways of getting involved in the rural areas.

The second publication is *We will find happiness in the rural areas* by Kazumasa Sashide, editor-in-chief of the magazine *Sotokoto* (Sashide 2016). Sashide pointed out that the number of mostly young people who find value in the "local," such as the rural areas, is increasing, and that "residence" is not necessarily an absolute condition in the relationship with the "local," but rather there is a variety of involvement styles. Sashide clearly defined "the population that is involved with the community" as the relationship population. Sashide also pointed out that local problems will not be solved even if we increase exchange population by promoting the attractiveness of the regions. Rather, it is more important to have people who can substantially and concretely solve problems by commuting to the area or communicating from remote locations, even if they do not reside in the area. Sashide claims that this is the true meaning of relationship population. And, in order to increase the relationship population, it is more significant to show "potential for relationship," not by showing the attractiveness of the region but rather its "weakness, cracks, and destruction." This "potential for relationship" will become the latent power of the region and the key to community development in the future, Sashide claimed.

The third publication is Terumi Tanaka's book, *Creating relationship populations* (Tanaka 2017), which was published in October 2017 and is considered the first book to include "relationship population" in its title. The book describes the background behind the emergence of the concept of relationship population and the role it plays in solving problems in both urban and rural areas. Specifically, the book uses the expression "hometown refugees" as the push factor for the out-migration from urban areas, pointing out that the anxiety of urban residents with no place to go leads to their longing for rural areas. Moreover, it mentions various forms of providing support to rural communities, which have contributed to solving the problems of the rural areas. The book also shows that residing in the area is not the only way to solve local problems, emphasizing a new relationship between urban areas and rural areas. It is also interesting that in the case of the residential population, involvement is only regarded for one area, while in the case of the relationship population, it is possible to become involved with several communities.

These publications offer a clear alternative to loudly advocating regional revitalization, where municipalities are forced to maintain a residential population based on a "population vision." Obviously, if the population of Japan as a whole is declining and its fertility rate continues to be low, maintaining residential populations may cause competition for migrants. On the other hand, the concept of exchange population, which presents various ways of involvement with communities, has become trivialized and equated with the volume of tourist arrivals. The February 2018 issue of *Sotokoto* magazine published a special feature titled "Introduction to relationship population," with the subtitle "More than tourism, less than migration." Thus, the relationship population is clearly not the same as the residential population or the exchange population but rather emerges as a population in between them.

Although the relationship population has been categorically defined by the series of publications that led to its conceptualization, this chapter further presents it is an even more diverse concept, in view of past policies and the reality of the regions. The following discussions attempt to systematically comprehend the concept of relationship population in its present stage.

5.2.2 Diversity in the Concept of Relationship Population

The government has officially presented the relationship population concept in a report compiled by the Ministry of Internal Affairs and Communications "Study Group on Future Immigration and Exchange Policies" (MIC 2017). This report points out that in order to maintain and strengthen the regional potential of rural areas, where population decline and aging are advancing ahead of other regions, it is important to increase the mobility of the population, e.g., through migration from and exchange with the urban areas, with a view to securing the players who will take the lead in the development of communities in diverse ways. It defines relationship population as "people who are variously involved with the communities and local residents and is different from either the long-term 'resident population' or the short-term 'exchange population,'" arguing for the need to increase and maintain the relationship population.

These assertions are believed to have stemmed from the following understanding of the relationship between urban areas and rural areas. In rural areas, it is necessary to maintain and strengthen regional potential population in order to solve problems related to the lives of residents, such as the degradation of the functions established for daily life support. Therefore, the report emphasizes the need to solve the shortage of people responsible for community development and bring in more human resources from outside the region by further promoting migration and exchange policies. This shows that the Ministry of Internal Affairs and Communications (MIC 2017) views relationship population from a quantitative perspective. This perspective is the reason why the relationship population is considered the "third population," which is different from either the residential population or the exchange population.

However, there are diverse human resources who support the rural areas from the outside. The first are those who have roots in the area, i.e., those who live close by in neighboring municipalities, as well as those who live far away in remote municipalities. Having people supporting the rural areas both from near and far suggests the reality and potential of supporting a community, even if one does not reside in the area. This is an epochal idea in that it points to the possibility of relying on human resources who can help the region without the premise of residing in the region. The report also mentions those who have no roots in the area. Specifically, they are those who are somehow involved with the community, like those who have worked, lived,

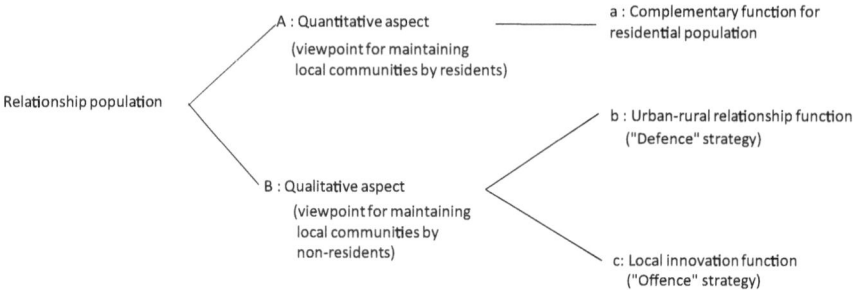

Fig. 5.1 Conceptual arrangement of relationship population

or stayed in the area in the past, or those who have visited the area due to business, leisure, or volunteer activities—the so-called wind-like persons.

Therefore, since there are diverse ways to get involved in the community, it is necessary to discuss relationship population from several angles. This chapter in particular presents two aspects of the relationship population, namely (A) the quantitative aspect and (B) the qualitative aspect (Fig. 5.1).

In the quantitative aspect (A), the relationship population is viewed in terms of the process leading up to the establishment of the residential population, where the strengthening of the relationship population ultimately leads to the maintenance of the residential population. In this case, the community cannot be maintained without a certain level of the residential population. Since it is based on the idea of the required population for community maintenance, the relationship population carries out a complementary function (a) for the residential population. On the other hand, in the qualitative aspect (B), the impact of the relationship population on the community is evaluated, rather than considering the significance of the relationship population only from a quantitative perspective. From the qualitative aspect, the relationship population can be viewed in terms of two functions, namely the urban-rural relationship (b) and local innovation (c). The urban-rural relationship function of the relationship population points to the fact that the maintenance of the community is clearly not a role exclusive to the residential population. This is a revolutionary concept that calls for a revision of the conventional understanding of the significance of residing in a region. On the other hand, the local innovation function of the relationship population brings about a softening of the rigid approach taken in the regions. This points to people who contribute to the maintenance of the community by stimulating and cooperating with the community—people who have been recently referred to as "wind-like persons."

These two functions from the qualitative aspect also respectively correspond to the "defense" and "offense" strategies in community development. In order to maintain the community, "defense" players are needed to maintain the regional functions, while "offense" players are needed to fulfill the potential of the community. The relationship population can contribute to both strategies.

The following sections examine the reality and significance of the relationship population by focusing on these three functions.

5.3 Relationship Population from the Perspective of the Complementary Function for Residential Population

5.3.1 Limitations of Maintaining Residential Population in a Society with a Declining Population

The quantitative aspect of the relationship population is based on the idea that only residents can take on the responsibility for the maintenance of the region. The concept of relationship population was born through shaking up this very idea. The following discussion examines the reality of residents being the ones responsible for maintaining the region and the process by which this situation has reached its limits.

There are three settlement functions in rural areas, namely the mutual support function, the production complement function, and the regional resources management function. Although these functions have been weakened due to the depopulation occurring during the high economic growth period, they have somehow been maintained until today by the remaining population. On the other hand, even in urban areas, regional autonomy has been maintained through community and neighborhood associations organized for each residential area.

As such, strongly cohesive local communities have been formed, both in urban and in rural areas. These have often been referred to as "village communities," and they have tended to exclude migrants and participants from outside the region, who were considered "outsiders." Strong local communities that tended to preserve the "village community" have even been considered "targets for eradication," since they resisted the formation of democratic communities during the high economic growth period.

Since then, depopulation has progressed in many rural areas, and even in urban areas, the number of people taking on the responsibility of maintaining the community has decreased as the population decreased. Consequently, it has become necessary to increase the residential population, where, instead of excluding "outsiders," these people were included among those actively invited to join the communities. Therefore, migration and residence have been raised as major policy issues in all municipalities, leading to a competition in attracting population.

However, since Japan has extremely few immigrants from overseas, even if domestic migration were promoted, the total population of Japan as a whole would not change, leading to the situation of a zero-sum game. The population attraction policies of each region have thus resulted in a "competition for population."

The relationship population may play a role in reversing this reality. In other words, it is possible to protect the community through various ways of involvement with it by breaking through the conventional mentality that only those who reside in the community are capable of protecting it.

5.3.2 Relationship Population to Complement the Required Population for Community Maintenance

Today, with the population declining in most regions, there is a shortage of personnel among organizations formed by residents, in which members are starting to feel weighed down by this burden. At first, they sought to increase exchange population by finding ways to complement the shortage of staff and by holding events to motivate themselves. However, the residents began to realize that there are many roles that only residents can play, no matter how much the exchange population increases.

Furthermore, due to the diversification of the values of local residents, some of those who live in the area do not participate in events and services to maintain the area. Some households have even declared that they will not affiliate themselves with the community association or administrative district, cutting themselves off from relationships with their neighbors, even though they live in the community. As such, in addition to the decrease in population and the aging of its members, there is a lack of residents involved in community maintenance activities, which has increased the burden on those who have continued to participate in these activities. Meanwhile, it is extremely difficult for communities to reduce the number of organizational positions and activities.

Some non-residents, however, continue to participate in the maintenance activities of their households and communities of origin. For example, some people who have moved out of the community due to marriage or employment frequently visit their parents' homes to help them, participate in events in the area where they are located, or serve on community boards. Some of them belong to the fire brigade and are indispensable members of their community. In describing this situation, Tokuno and Kashio (2014) emphasized the significance of their role as modified extended families.

Therefore, although it is possible that various stakeholders aside from residents can play a role in maintaining the community, there is still a minimum necessary number of residents. In this paper, the author refers to this number as the required population for community maintenance. The above situation provides the context for how the relationship population performs the function of complementing the residential population.

5.4   New Urban-Rural Relationships from the Perspective of Relationship Population

5.4.1   Significance of the Existence of Rural Areas from the Perspective of Urban Areas and Their Relationship

Recently, due to the diversification of lifestyles and the progress of the internet society, especially among young people, workstyles and workplaces have also changed significantly. Furthermore, after the Great East Japan Earthquake, there has been an increasing tendency to emphasize "social value" as embodied by an "inclination to be useful to people." As a result, an increasing number of urban residents are eager to move to rural areas. According to a survey by the Ministry of Internal Affairs and Communications (MIC 2018), 30.6% of urban residents hope to migrate to the countryside. In particular, those in their 20 and 30s tended to have a high inclination to move to rural areas.

In addition, many of the urban residents desire to be involved in rural areas by means other than relocating, and they are interested in sightseeing and participating in events. Around 10% of them either wish to stay in order to participate in local activities and interact with local people or have residences in both regions. Furthermore, by age, the proportion of residents in their 20 and 30s who wish to make a U-turn to their municipalities of origin exceeded 40% of all age groups. Therefore, although many urban residents wish to relocate or make a U-turn to rural areas, many of them are unable to readily do so.

In other words, although many urban residents want to move to rural areas, or to become involved with them, the hurdle for doing so is high. Therefore, it is necessary to focus on the relationship population, or those who are variously involved with the community and with local residents but are not part of the long-term resident population or the short-term exchange population. This points to the importance of forming a continuous and multi-layered network with human resources from outside who are interested in a particular community, and to deepen and sustain the involvement of those who contribute to the community as members of the relationship population.

The provision of funds, knowhow, and labor by human resources from outside a region is easy to link to intrinsic movements within the region. Accordingly, it is important to deepen the relationship with the community through migration and exchange and to carry out autonomous and continuous community development through cooperation with stakeholders within and outside the community.

Thus, the concept of relationship population is significant in that it has changed the conventional dichotomy between urban and rural. In other words, it can be said that the two are mutually complementary rather than conflicting entities. Moreover, urban residents and rural residents can form a variety of relationships.

5.4.2 Diverse Urban-Rural Relationships

Figure 5.2 shows the types of relationship populations in terms of the involvement between urban areas and rural areas. Thus far, the relationship population has tended to be viewed only from the perspective of urban areas. As a result, the flow from the exchange population to the residential population has become fixed, and it has been presented as though it were not possible to flow in the opposite direction or to stop at each step. Therefore, in Fig. 5.2, the perspective from the rural area was added along the vertical axis. The objectives of the relationship population, meanwhile, are diverse. The figure presents two axes, namely "emphasis on life maintenance," which focuses on the "defense" aspect, and "emphasis on value creation," which focuses on the "offense" aspect.

Therefore, the relationship population can be classified into the following four types of people. Regional support-oriented people (A) aim to solve local problems and utilize local resources by creating new values from the perspective of the rural areas. Among the four types, they are the most actively involved in rural areas, and by extension, most seriously consider migration to the community. Slow-life-oriented people (B), on the other hand, recognize the value of rural areas and endeavor to cherish that value, but they do so from the perspective of urban areas. Therefore, they have little interest in the daily-life issues of rural areas. Regional-contribution-oriented people (C), like those of type (B), view things from the perspective of urban areas, but they are highly aware of the problems faced by rural areas, and they make every effort to contribute to the solution of those problems. They have a strong sense of their role as the so-called community support team, but unlike type (A) people, they do not consider the possibility of migration. Non-residential-area-maintenance people (D) are fully aware of the problems faced by rural areas and are strongly committed to protecting the livelihoods of the community. In many cases, they are either children or grandchildren of people from the community or are those who have moved out of the area.

Thus far, relationship population has been viewed as comprising either regional-support-oriented (A) type or regional-contribution-oriented (C) type of people. Many

Fig. 5.2 Relationship population from viewpoint of urban-rural relations. *Notes* Drawn mainly based on Taguchi (2018)

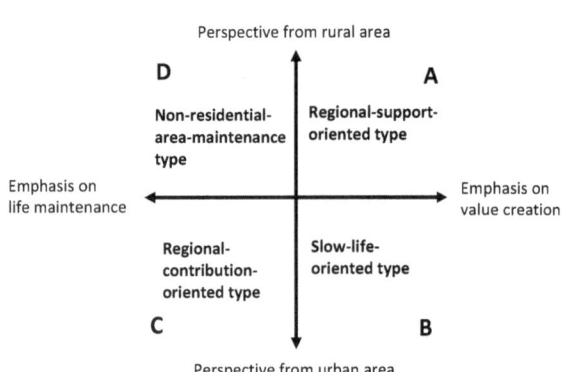

of the relationship population policies are along the A and C axes. Examples include the hometown tax, hometown resident certificate system, hometown volunteer brigade, and hometown working holidays. Therefore, types A and C can be referred to as the "relationship population in the narrow sense."

On the other hand, the non-residential-area-maintenance (D) type includes mostly out-migrants, and it has not been mentioned in discussions on relationship population thus far. But in reality, these people still have a strong relationship with the community and are actively involved in its maintenance. Sakuma et al. (2016) recognized the significance of their roles early on, referring to them as "external family" or "quasi-family." Moreover, as Tokuno and Kashio (2014) claimed, they are referred to in sociology as the modified extended family in discussions on the mechanisms of community maintenance. This type of people should be recognized as the "relationship population in the broad sense."

Meanwhile, the slow-life-oriented (B) people, despite being aware of the limitations of urban areas and the sustainability of rural areas, are not willing to leave their urban areas. Given the success of events held in rural areas and regional fairs held in urban areas, we can say that the majority of those who have an appreciation for the rural areas belong to this category.

5.4.3 Policies Relevant to Relationship Population Aimed at Building New Urban-Rural Relationships

Table 5.1 shows examples of relevant national and municipal measures that traverse the relationships between the four types of relationship populations and the three functions they perform as discussed above. The "number of projects" in the table shows the number of projects adopted as model projects in the government's "Relationship Population Creation Project."

According to the table, the regional support-oriented type has a high affinity with the local innovation function (discussed below). National projects such as the Trial Satellite Office and Telework in Hometown, as well as the Hometown Supporter System (Nanto City) and Platinum Residence (Kimotsuki Town) fall under this category. In addition, there are also many projects related to the urban-rural relationship function, including the Project for Interaction between Urban and Rural Areas, Hometown Resident Certificate System (Hino Town, Miki Town, etc.), and Hometown Foundation (Saijyo City, etc.). In all, 13 relevant model projects are regarded by municipalities as suitable for the development of relationship population.

The slow-life-oriented type (B) of relationship population embodies the urban-rural relationship functions. For example, the Hometown Tax and Hometown Supporter System (Shibata City, Uonuma City) projects are characterized by a loose connection with their related areas. They can be considered relationship population projects involving people who, despite living in urban areas, are inclined to widely support the rural areas.

Table 5.1 The function of relationship population and the projects by National government and local municipality by each type of secure relationship population

Type of relationship population	A : Regional-support-oriented type	B : Slow-life-oriented type	C : Regional-contribution-oriented type	D : Non-residential-area-maintenance type	Number of project
Standpoint	Rural area	Urban area	Urban area	Rural area	
Important element	Value creation	Value creation	Life maintenance	Life maintenance	
Function of relationship population					
a : Complementary function for residential population			Working holiday in hometown	Modified extended family	2
			Information center for relationship population (Mima City, Sanagochi Village, Minami Town)		
b : Urban–rural relationship function	Project of interaction between urban and rural areas	Hometown tax donation program	Hometown supporter club (Tomakomai City, etc.)		20
	Hometown resident certificate system (Hino Town, Miki Town)	Hometown supporter (Shibata City, Uonuma City)	School for hometown supporter (Yokote City)		
	Hometown foundation (Saijyo City, etc.)		Hometown supporter member (Amakusa City)		
c : Local innovation function	Trial satellite office				8
	Telework in hometown				
	Hometown supporter system (Nanto City)				
	Platinum residence (Kimotsuki Town)				

(continued)

Table 5.1 (continued)

Type of relationship population	A : Regional-support-oriented type	B : Slow-life-oriented type	C : Regional-contribution-oriented type	D : Non-residential-area-maintenance type	Number of project
Standpoint	Rural area	Urban area	Urban area	Rural area	
Important element	Value creation	Value creation	Life maintenance	Life maintenance	
Number of project	13	7	10	0	30

Source This tabulation is mainly based on the application form of relationship population creation project by Ministry of Internal Affairs and Communications (2018)

Note "Number of projects" in the table indicates the number in all model projects by the ministry

Likewise, regional-contribution-oriented (C) people also embody the urban-rural relationship functions. The Hometown Supporter Club (Tomakomai City, etc.), School for Hometown Supporter (Yokote City), and Hometown Supporter Member (Amakusa City) projects are more oriented toward the relationship with the target area than the slow-life-oriented (B) relationship population. This type also embodies the complementary function for residential population. It includes people who are deeply involved in the target communities through the Working Holiday in the Hometown project and those who aim to establish strong future involvement through the Information Center for Relationship Population project (Mima City, Sanagouchi Village, and Minami Town).

Among the model projects adopted, 20 projects are classified under urban-rural relationship functions, and they account for two-thirds of the total. Following these, there are several (eight) projects classified under the local innovation function but only two projects classified under the complementary function for residential population. As mentioned above, it is assumed that the function of complementing the residential population, a function that the relationship population exhibits, is important in maintaining a region's population. However, it is clear that the recognition of this function is low, even among local governments.

5.5 Significance of Relationship Population in Bringing About Local Innovation

5.5.1 Relationship Between Exchange Population and Relationship Population

In discussions of relationship population, its difference from the exchange population usually comes into play. What we call relationship population may have originally been included in the concept of exchange population. In this section, we reexamine the relationship between exchange population and relationship population.

According to Hirao (2003), the concept of exchange population was first used by the Nihon Keizai Shimbun newspaper in 1990.[5] After that, in 1994, the National Land Agency proposed the concept of exchange population as an index for promoting exchange policies. National Land Agency, Planning and Coordination Bureau (1994) defined exchange as "a relationship that has some effect on the region, whether or not there is a visit to the region." In addition, the "exchange population serves as an indicator for systematically grasping the effects of exchanges and their impact on local communities." In this definition, the exchange population also includes those who "do not visit" but are involved in a way that has some effect on the community.

Thus, the exchange population is essentially seen as a population that complements the residential population, and conceptually it includes not only the visitors to the area but also those who have some kind of involvement with the community. Since this corresponds directly to the relationship population, it is clear that exchange population includes the current concept of relationship population. However, since appropriate data could not be collected for measuring the exchange population, other collectable data, such as data on tourists and guests, have been gathered instead.

Measuring the exchange population based on the number of tourists and hotel guests is largely attributed to efforts taken to revitalize the local economy. For example, according to the 2015 MIC White Paper on Information and Telecommunications, the annual per capita consumption for the residential population (1.24 million yen) is equivalent to that for 10 foreign travelers, 26 domestic overnight travelers, or 83 domestic day-trippers. This idea of equivalence considers the residential population only as economic consumers, and thus, in the same manner, the exchange population is also valued only in terms of how they complement the residential population economically.

Therefore, due to the lack of measurement indices and the overemphasis on the economic significance of the exchange population, which is assumed to conceptually include the relationship population, the impact of the exchange population and the support it provides to the community had not been properly taken into account.

5.5.2 Significance of Local Innovation and the Role of Relationship Population

Since the 1960s, depopulation has continued in the rural areas, and a sense of resignation has pervaded many regions. Today, in an era of declining population, the population in urban areas, which had until recently been on an increasing trend, is now also on a declining trend. Due to a vague apprehension about the future and the inability to find appropriate countermeasures, pessimism toward the regions has prevailed. Many experts have pointed out that "restoration of confidence and pride" is important in such regions (e.g., Odagiri 2014; Inagaki et al. 2014). Although local residents should play the main part in such initiatives, it is difficult for them alone to carry out creative community development to create a new value system in the region (Taguchi 2018).

Consequently, this points to the need for those who are outside the region to act as external players. By assessing the value of a region from an external perspective, external players can promote awareness of the region's value within the community. In addition, the involvement of external players will likely encourage local residents and elicit fresh initiatives within the community. This will lead to the discovery of new possibilities for the community from the "defense" aspect to solve local problems, as well as from the "offense" aspect to optimize use of local resources. In this chapter, this series of activities is regarded as local innovation, and the relationship population is believed to fulfill the role of external players mentioned above. In the following, we discuss how the relationship population is involved in local innovation from the standpoints of "defense" and "offense."

First, from the standpoint of "defense," the players can be divided into C and D as shown in Fig. 5.2. In particular, those who fall under the non-residential-area-maintenance type (D) are originally from the area and have moved out of it or, sometimes, their grandchildren. Most of them regularly come to visit their family members who continue to live in the area. They contribute to the maintenance of local households, such as by shopping for their families, sending them to hospitals, and helping them with farm work. By extension, they have become indispensable for community maintenance activities through such participation as attending community meetings, helping to clear grass areas and clean waterways, and assisting in festivals and fire brigade activities.

They are recognized as quasi-members of the community, even though they obviously do not reside in the area. Their roles have not been sufficiently recognized in the past because they exist as a normal part of every community. In the future, there is a need to clearly recognize their role as members of a relationship population that supports the maintenance of the community.

There is also potential for local innovation from the standpoint of "offense." The main player for local innovation is the regional support-oriented (A) relationship population, as shown in Fig. 5.2. Although they live mainly in urban areas and earn a stable income, they have doubts and anxieties about living in these urban areas. At the same time, however, it is difficult for them to migrate to rural areas. According to Sashide (2016), they endeavor to achieve their aspirations in the rural areas around keywords such as "local" and "social." Therefore, they create opportunities to do this by proactively going out and participating in events in rural areas, and they continue to maintain relationships with the community through social media and other means.

Some of them may have a local living base because they frequently travel to and from the region in which they have become involved. Depending on the degree of their involvement in the community, some of them belong to local community or neighborhood associations and function as the non-residential-area-maintenance (D) type of relationship population. This type of involvement is similar to the conventional arrangement of having two places of residence, or two local bases, and thus these arrangements can also be incorporated in the concept of relationship population.

5.6   Conclusion

In this paper, we reexamined the concept of relationship population and presented a new perspective on its significance. Thus far, the relationship population has been regarded as "the third population concept intermediate between the exchange population and the residential population." Although the author does not deny such an intermediate nature of the relationship population, through this study, the author has come to the conclusion that it is misleading to explain these three population concepts only in terms of stages. In other words, we should consider the relationship population not only as a step between the exchange population and the residential population but also as one of the modes of involvement between urban areas and rural areas in the new era.

There is also a diversity of relationship populations. In this paper, the author categorized relationship population into four types based on the relationship between urban and rural areas and summarized the characteristics of each type. In past discourse on relationship population, much attention has been given to the regional support-oriented type and the regional-contribution-oriented type of relationship populations. However, maintenance of the region necessitates human resources who can continuously play an active role in protecting the region. In this paper, the author showed that the non-residential-area-maintenance type of relationship population provides the human resources to perform such a role. Specifically, the concept of the modified extended family, which has been forwarded by Tokuno and Kashio (2014), should be clearly regarded as part of the relationship population.

In the future, the concept of relationship population will likely change significantly as it is examined in various fields, and thus we must keep a close eye on these developments.

Notes

1. The Community-Reactivating Cooperator Squad (CRCS) program requires the transfer of resident certificates from urban areas to disadvantaged regions such as depopulated areas. While living there for a certain period of time, the participants engage in "community collaboration activities," such as supporting local revitalization, engaging in agriculture, forestry, and fisheries, and supporting the livelihoods of residents, while living and establishing themselves in the area. As of the end of FY2021, 6015 people are conducting activities in 1085 municipalities nationwide under the CRCS program.
2. Miyaguchi (2007) defines community development as "the process of intrinsically creating local values suitable for the times and adding them to the community." The discussions in this paper basically follow this idea.
3. Regional Management Organization is a concept proposed by the Ministry of Internal Affairs and Communications in 2012. The mergers of municipalities around 2004 led to the establishment of new self-governing resident organizations. Regional Management Organizations, in particular, carry out executive functions and are formed to solve local problems and utilize local resources. As of FY2021, there are 6064 such organizations nationwide.
4. U-turn is a term commonly used in Japan to refer to returning to one's place of origin, while I-turn is used to refer to moving to a place other than one's place of origin.
5. The Nihon Keizai Shimbun Local Economy Page in the January 17, 1990, issue reported that when the construction of a golf course in Asahikawa City started, the Chairman of the Asahikawa Chamber of Commerce and Industry said, "the decrease in the residential population in Asahikawa needs to be compensated for by the increase in the exchange population".

References[1]

Hirao M (2003) Koryu sangyo to chiiki keizai no seicho (Exchange industry and growth of regional economy). Chiiki Keizai Kenkyu (Regional Economic Studies),14: 37–50.(J)

Inagaki F, Abe I, Kaneko T. et al (2014) Shinsai hukko ga kataru nosanson saisei (Rural regeneration in terms of post-earthquake reconstruction). Komonzu, Tokyo.(J)

Ministry of Internal Affairs and Communications (2017) Korekara no iju koryu sesaku no arikata ni kansuru kentokai chukan torimatome (Interim report of policy for migration and exchange population by the Investigative Commission). Ministry of Internal Affairs and Communications, Tokyo.(J)

Ministry of Internal Affairs and Communications (2018) "Den'en kaiki"ni kansuru chosa kenkyu hokokusho (Research report on "migration to the countryside"). Depopulation Countermeasures Office, Regional Power Creation Group, Ministry of Internal Affairs and Communications, Tokyo.(J)

Miyaguchi T (2007) Shin chiiki wo ikasu: Chiri gakusya no chiiki zukuri ron (Making the most of the region (new edition): Geographers' view of regional development). Harashobo, Tokyo.(J)

National Land Agency, Planning and Coordination Bureau (ed.) (1994) Koryu jinko: Aratana chiiki seisaku (Exchange population: New regional policy). Ministry of Finance, Printing Bureau, Tokyo.(J)

Odagiri T (2014) Nosanson wa shometsu shinai (Rural areas will not disappear). Iwanami Shoten, Tokyo.(J)

Sakuma Y, Tsutsui K, Kiboishima H (2016) Nosanson shuraku no seikatsu wo sasaeru chiiki gai kazoku no nettowaku no jyokyo ni kansuru chosa: Aichiken Kita-shitara gun Toyone mura ni okeru 2002nen chosa to 2015nen chosa no hikaku wo tsujite (Investigation on the status of networks of out-of-region families that support the lives of rural villages in 2002 and 2015: Case of Toyone Village, Kita-shitara Gu, Aichi Prefecture, Japan). Journal of San'en Nanshin region cooperation center of Aichi University, 4:25–32.(J)

Sashide K (2016) Bokura wa chiho de siawase wo mitsukeru (We find happiness in the rural areas). Popura-sha, Tokyo.(J)

Taguchi T (2018) "Kankei jinko" no chiiki zukuri niokeru kanosei (Possibility of relationship population for community development). Chosa Kenkyu Johoshi (Journal of Ehime Center for Policy Research), 2:13–18. (J)

Takahashi H (2016) Toshi to chiho wo kakimazeru: "Taberu tsushin"no kiseki (Mixing urban and peripheral areas: Miracle of "eating communication"). Kobunsha, Tokyo.(J)

Tanaka T (2017) Kankei jinko wo tsukuru: Teiju demo koryu demo nai rokaru inobesyon (Creating relationship populations: Local innovation, neither residence nor exchange). Kirakusya, Tokyo.(J)

Tokuno S, Kashio T (2014) Kazoku shuraku josei no sokojikara: Genkai shuraku-ron o koete (Potentiality of family, community and women: Beyond the marginal settlement discourse). Nosangyoson Bunka Kyokai, Tokyo.(J)

---

[1] (J): written in Japanese